D1826495

The Cream of Strawberry Shortcake Collectibles

Jan Lindenberger

with Jennifer Bowles

4880 Lower Valley Road, Atglen, PA 19310 USA

Copyright © 1999 by Jan Lindenberger
Library of Congress Catalog
Card Number: 98-89926

All rights reserved. No part of this work may be reproduced or used in any form or by any means—graphic, electronic, or mechanical, including photocopying or information storage and retrieval systems—without written permission from the copyright holder.
"Schiffer," "Schiffer Publishing Ltd. & Design," and the "Design of pen and ink well" are registered trademarks of Schiffer Publishing Ltd.

Book Design by Anne Davidsen
Type set in Shelley Allegro/ Korinna

ISBN: 0-7643-0812-2
Printed in China
1 2 3 4

Published by Schiffer Publishing Ltd.
4880 Lower Valley Road
Atglen, PA 19310
Phone: (610) 593-1777; Fax: (610) 593-2002
E-mail: Schifferbk@aol.com
Please visit our web site catalog at
www.schifferbooks.com

This book may be purchased
from the publisher.
Include $3.95 for shipping.
Please try your bookstore first.
We are interested in hearing from authors
with book ideas on related subjects.
You may write for a free catalog.

In Europe, Schiffer books are distributed by
Bushwood Books
6 Marksbury Rd.
Kew Gardens
Surrey TW9 4JF England
Phone: 44 (0)181 392-8585
Fax: 44 (0)181 392-9876
E-mail: Bushwd@aol.com

Contents

Acknowledgments

A very special thanks to Jennifer Bowles for her patience and kindness, for welcoming me into her home and allowing me to photograph her vast collection. Jennifer basically contributed all the valuable information for this book . She would like to be in touch with other avid Strawberry Shortcake collectors. Jennifer publishes the *Strawberryland Gazette,* a newsletter about Strawberry Shortcake and her friends, and welcomes any correspondence from other collectors. Jennifer buys and sells as well. For information on her newsletter (that is published 6 times a year, $24.00 yearly or $4.00 an issue) you may contact her at:

Strawberryland Gazette
138 E. Main Cross
Greenville, Kentucky 42345

or by phone: 1–502–338–4318

Also, thanks to Helen Bowles for all her help with arranging and re-arranging her daughter Jennifer's vast collection and for those wonderful home-cooked meals.

And to:

Linda Strumski, Waterbury, Ct.

Michelle Razzi, Baltimore, Md

Berri Bits
C/O Pat Minks
10001 Elliott Ave S
Bloomington, MN 55420

Gayle Anderson,Cambridge, Mass

Jennifer Bowles has loved and collected dolls all her life (26 years)! Jennifer works at an elementary school as a Speech Language Pathologist and runs a part-time business called The Doll Patch. She grew up loving Strawberry Shortcake and has been re-collecting Strawberry Shortcake for the past 5 years.

Introduction

If you are reading this, then you must have gotten the "Berry Bug" or are just curious! Strawberry Shortcake was created by American Greetings in 1979 and eventually became a line of dolls, miniatures, and accessories by Kenner from 1980–1985. Please refer to *More Strawberry Shortcake*, also available from Schiffer Publishing, for more information on these. However, like any toy craze, there were other licensed products besides the toys to perpetuate the madness! The many, many surfaces with Strawberry Shortcake's picture on them are now cherished memorabilia.

Now that previous books have created a foundation of knowledge about Strawberry Shortcake collectibles, this book will go a step further and portray some harder-to-find rarities that were produced on a limited basis and distributed for a shorter time than the mass-produced line of dolls and toys. Breakables like porcelain items and fragile ornaments are hard to find in mint condition, but a surprisingly wide assortment exists in the hands of collectors or the attics of grown-ups who loved Strawberry Shortcake when they were younger. An exciting aspect of collecting Strawberry Shortcake is having that hope of finding a rare item at a yard sale—and this does happen! A berrykin pet might peek out from a pile of old action figures or McDonald's toys, and the price might be a dime or a quarter. Quite a find when you consider its $50 value to a collector!

A big part of Strawberry's skyrocketing popularity is the nostalgia felt by previous collectors as they get old enough now to have jobs that can support a growing collection. A common wistful complaint is, "my mom threw out my dolls and now I'm going to buy them all again!"

Collecting Strawberry Shortcake in a serious way is a commitment to a certain way of thinking, an outlook on life that transcends the acquisition of the actual plastic dolls and toys. Strawberryland is a real place to us, a place without war and crime, without the big and little setbacks of life. Although we are now grown, we can be children again without apology when we find Lemon Meringue's hat or Plum Puddin's shoes or a Berry Merry Bug in a flea market discard box.

Strawberry Shortcake is an ideal, a philosophy of life that might be looked down upon smugly by collectors of more supposedly sophisticated dolls. But Strawberry's charm is in her very lack of sophistication, her refusal to be forced into the grown-up shoes of somebody else's idea of fashion and desirability. She is Peter Pan and she is Dorothy, forever young and innocent, forever ensconced in the Strawberryland of our girlhood dreams.

1984 Miniature, Peach Blush with fan. $75 and up loose, $100 and up MIP.

Satin ball ornament. $15–20 MIP

1983 Baby's First Christmas satin ball ornament. $10–15 MIB

1982 Satin ball ornament. $10–15

Satin ball ornament. $10–15

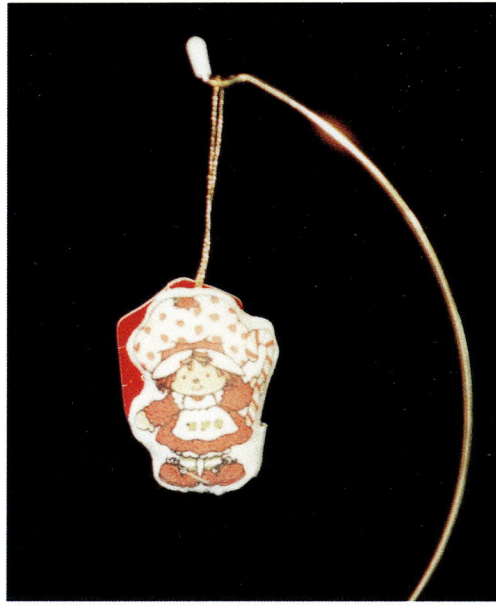

Small cloth ornament. $10–15

Satin ball "Cozy Christmas" ornament. $15–20

Ceramic ornament: Strawberry Shortcake giving Custard a candy cane. $20–25

Resin ornament: Strawberry Shortcake holding present. $20–25

Strawberry Shortcake ornament: "A Special Gift". $45 and up MIB

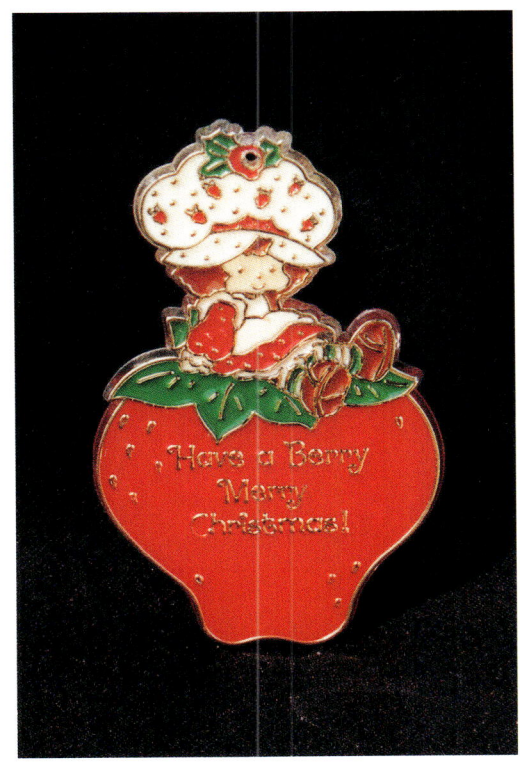

Plastic ornament. $15–20

Plastic panoramic view ornament. $30–up

Strawberry Shortcake ornament: "Up, Up and Away". $75 and up MIB.

Christmas snow dome. $20–25 MIB

Ceramic ornaments. Apple Dumpling on sled, Custard in stocking, Strawberry Shortcake looking through wreath, Custard in present. $25–35 each

Ceramic ornament: Strawberry Shortcake holding a star. $20–25

Ceramic pink bell-type ornament. $15–20

Ceramic ornament: Strawberry Shortcake holding a star. $15–20

Ceramic ornament: Raspberry Tart with Christmas tree. $35 and up

Ceramic ornament: Strawberry Shortcake holding a berry with Custard. $25–35

Ceramic ornament: Strawberry Shortcake with Custard. $25–35

Ceramic ornament: Strawberry Shortcake with tree. $25–35

Ceramic ornament: Apple Dumpling with Custard. $10–15

Ceramic ornament: Raspberry Tart with Candy Cane. $15–20

Ceramic ornament: Strawberry house. $20–25

Ceramic ornament: Strawberry Shortcake on a star. $20–25

Ceramic ornament: Strawberry Shortcake with cake. $20–25

Ornament: Apple Dumpling sitting in a star. $45 and up MIB

Ceramic ornament: Strawberry Shortcake holding large berry. $20–25

Ceramic ornament: Apricot on horse. $45 and up MIB

Ceramic ornament: Huckleberry Pie with drum. $45 and up MIB

Ceramic ornament: Strawberry Shortcake with snowman. $25–35

Porcelain ornament. $40–50

Porcelain Items

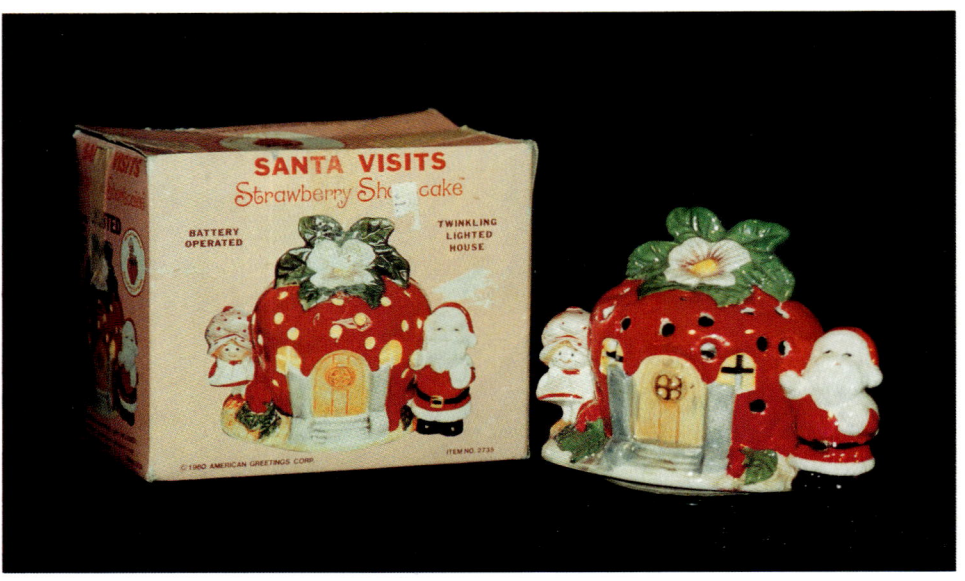

Ceramic light-up house of Strawberry Shortcake with Santa. Large size, $50–75 MIB. Small size, $35–45 MIB.

Ceramic light-up house of Strawberry
Shortcake berry house. Large size, $50–75
MIB. Small size, $35–45 MIB.

Ceramic light-up Strawberry Shortcake
house with pets. 2 sizes. $35 and up

17

Ceramic light-up Strawberry Shortcake with Santa. 2 sizes. $35 and up

Ceramic light-up Christmas tree. $35–40 MIB

Figural head mugs. Blueberry Muffin, Strawberry Shortcake, Raspberry Tart. $20–up each

Child's 3-piece porcelain dinner set. $40 and up

Ceramic coffee cups. $15–20 each.

Heart-shaped porcelain snack set with tags. $40–50

Porcelain cup and saucer set. $25–35

Stoneware mug: Strawberry Shortcake on a berry. $15–20

Ceramic and iron trivet. $20–up.

Stoneware Huckleberry and Strawberry mug. $15–20

Strawberry Shortcake "You're the Berries" porcelain mug. $20–25

Above: Ceramic and pink iron trivet. $25–up.

Right: Ceramic and metal letter holder. $50–60

Porcelain berry-shaped plate.
$15–20

Porcelain candy dish. $20–25

Porcelain berry-shaped plate.
$12–18

Porcelain plate. $10–15

Small Mother's day plate. $10–15

Heart-shaped plate. $15–20

Strawberry-shaped porcelain dish. $15–20

Porcelain pin cushion and thimble set. $40–50

Strawberry-shaped porcelain dish. $15–20

Porcelain toothbrush holder. $40 and up

Porcelain trinket box. $20–25

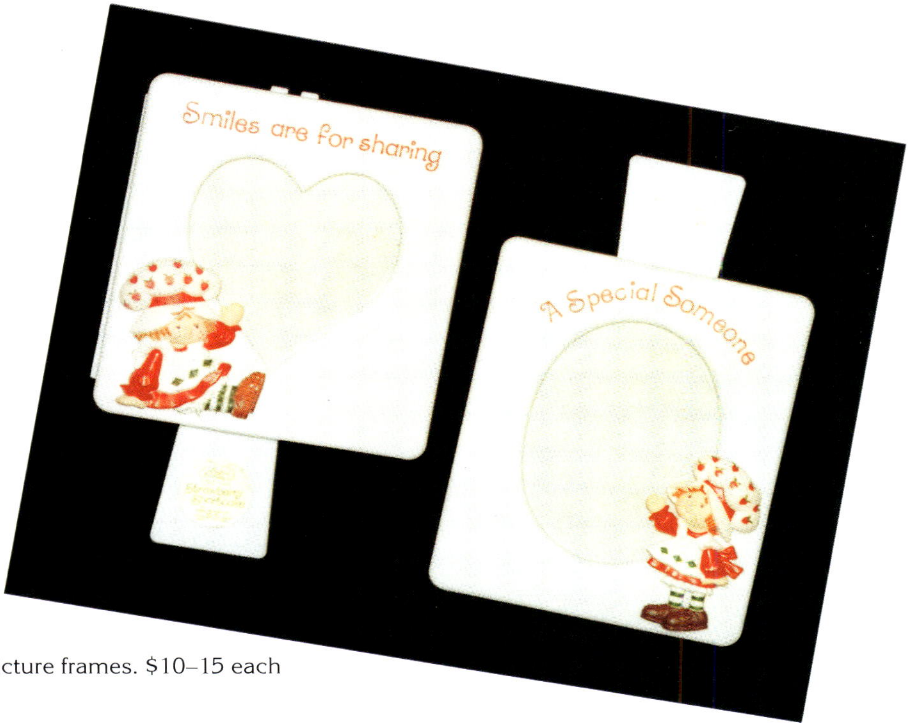
Smiles are for sharing
A Special Someone

Plastic picture frames. $10–15 each

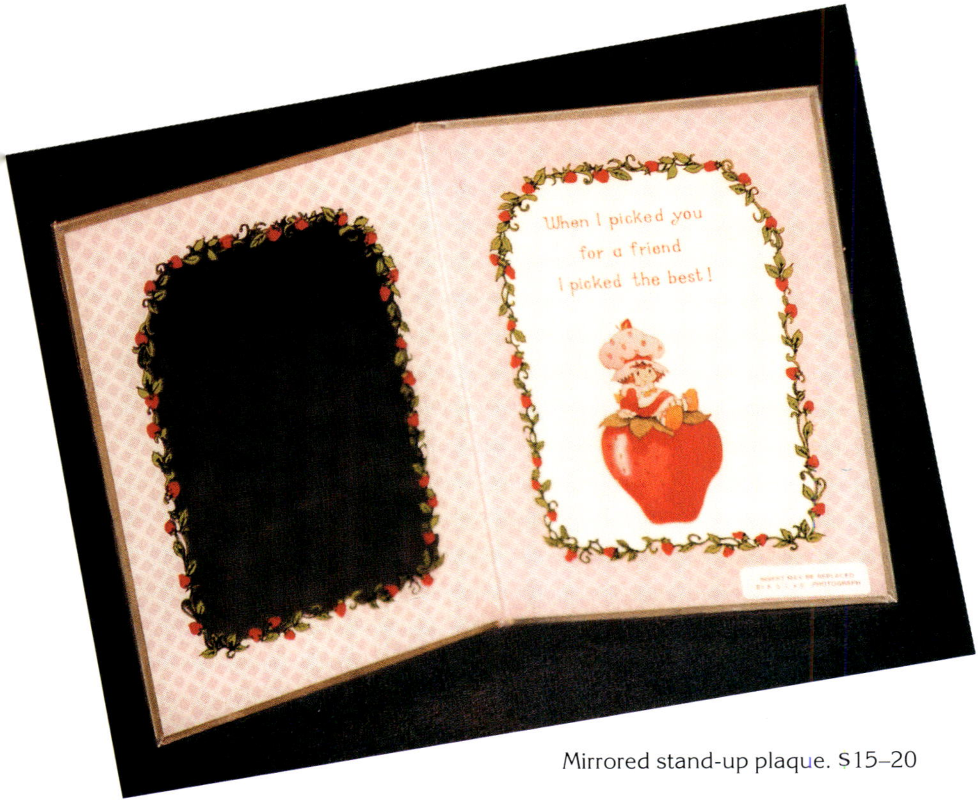
When I picked you
for a friend
I picked the best!

Mirrored stand-up plaque. $15–20

Silver plated bank. $15–20

Above: Ceramic moon bank.
$50 and up

Right: Huckleberry Pie on a
moon bank. Compo/plaster
material. American Greetings.
1983. $50 and up

Ceramic bank. $15–20

Ceramic figural mirror. $60–80

Ceramic pomander.
$12–15

Mother's day vase. $20–25 MIP

Valentine mobile.
$15–20 MIP

Small porcelain vases. $15–20 each

Porcelain vase. $15–20

Candles in gift boxes. $10–15

Crystallized candle holders. $20–25

Crysta lized candle holders. $20–25

Small porcelain candle holders. $15–20

Above: Large ceramic candle holder. $22–28

Left: Crystallized large Christmas candle. $20–25

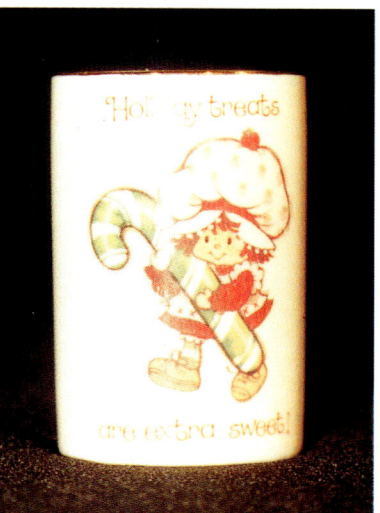

Above: Large porcelain candle holder. $20–25

Left: Holiday candle holder. $15–20

Plastic heart-shaped trinket boxes. $15–20

Plastic daisy-shaped trinket boxes. $15–20

Porcelain heart-shaped trinket box. $15–20

Porcelain heart-shaped trinket box. $15–20

Porcelain round trinket box. $20–25

Bisque figural trinket box. $20–25

Bisque trinket box. $20–25

Above: Porcelain round trinket box. $25–35

Left:Porcelain trinket box. $20–25

Porcelain sugar (Love is the treat that's extra sweet) and creamer(Pour a little kindness) service. Also available: jam jar similar to sugar jar that depicts "spread". $50–65 each

Above: Sugar jar. $40–50

Right: Porcelain spoon rest. $30 and up

Holiday porcelain bell. $20–25 MIB. $10–15 loose.

Porcelain bell. $15–20

Porcelain bell. $15–20

Porcelain bell. $15–20

Above: Porcelain sitting figurine. $20–25

Left: Porcelain bell with heart handle. $15–20

Below: Figural candles. $5–10 each

Porcelain figurines. (Some figurines have pink noses). $20–25 each

Ceramic figurines. Apple Dumpling, Huckleberry Pie, Strawberry Shortcake holding strawberry $15–25 each

Ceramic figurines. Blueberry Muffin, Raspberry Tart, Plum Puddin'. $15–25 each

Ceramic numeral cake toppers. Set of 8 with tags. $20–25 each

Ceramic numeral cake toppers. $20–25 each

Ceramic birthday numeral cake toppers. $15–20 each

Jewelry

Strawberry Shortcake holding Custard, ring.
$10–15 MOC

Strawberry Shortcake holding Custard,
necklace. $10–15 MOC

Strawberry Shortcake holding Custard, pin.
$10–15 MOC

Blueberry Muffin pin. $10–15

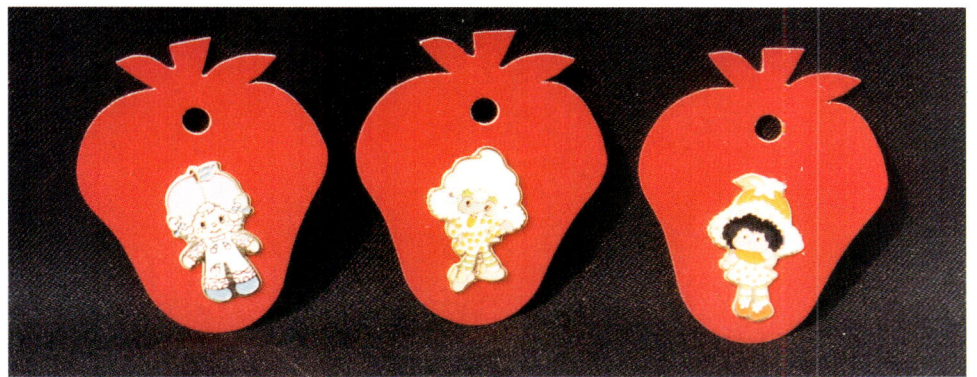

Pins: Apricot, Lemon Meringue, Orange Blossom. $10–15 each MOC

Pins: Strawberry Shortcake, Apple Dumpling, Blueberry Muffin. $10–15 MOC

Above: Earrings. $10–15

Left: Pin and bracelet. $5–10 each

Imitation pins. $5–8 each

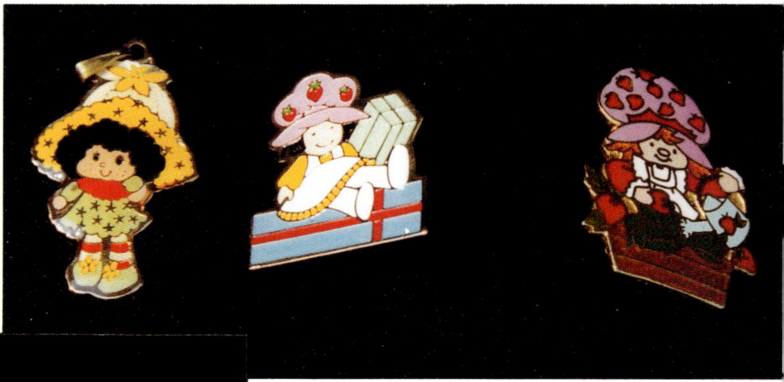

Above: Orange Blossom pendant. $10–15.
Imitation pin. $4–6. Strawberry Shortcake
at window box pin. $5–10

Left: Imitation key ring. $5–10

Below: Orange Blossom pendant. $10–15.
Imitation holiday pin. $5–8. Strawberry
Shortcake pin. $5–10

Package for 14kt. gold charm. $25–up with charm.

Strawberry Shortcake button. $5–10

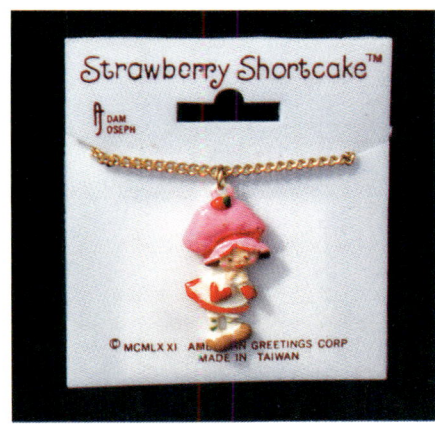

Strawberry Shortcake 3–D necklace. $10–15 MOC

Large plastic pendant with cord. $15–20

Plastic charm. $5–10

Mom's pin. $10–15

Ceramic pins. $10–15 each

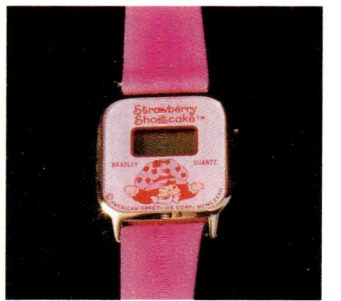

Digital pink watch. $15–20

Child's digital watch. $15–20

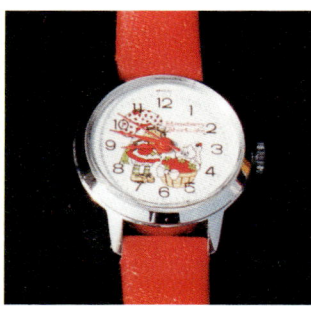

Child's watch $15–20

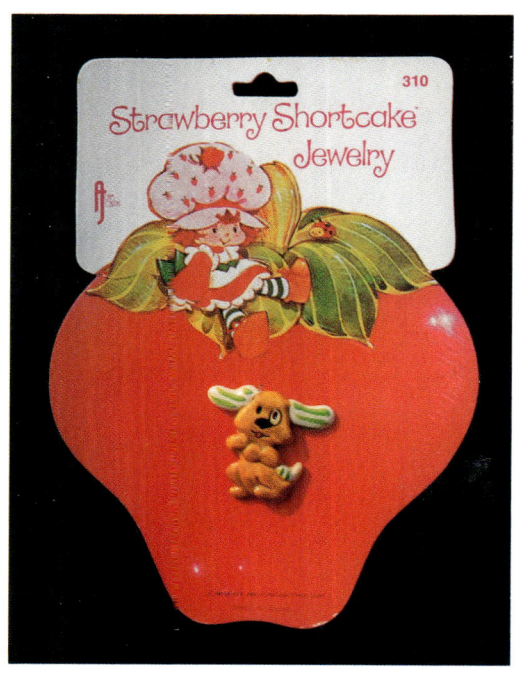

Pupcake pin. $10–15 MOC

Plastic 3–D Strawberry Shortcake pendant. $10–15 MIB

Plastic 3–D Apple Dumpling pendant.
$10–15 MIB

Plastic 3–D Apricot pendant. $10–15 MIP

Butter Cookie pin/pendant. $10–15 MIP

Cherry Cuddler pin/pendant. $10–15 MIP

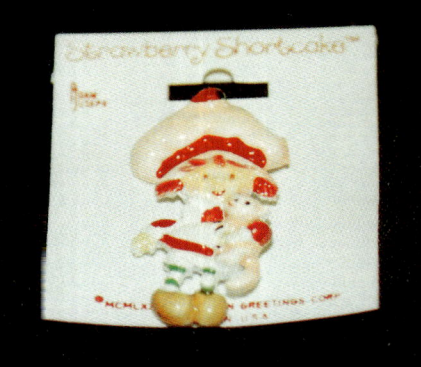

Metal 3–D pins. $10–15

Strawberry Shortcake pin/pendant. $10–15 MIP

Metal 3–D necklace. Strawberry Shortcake holding a berry. $10–15

Above: Strawberry Shortcake 3–D necklace. $10–15 MOC

Left: Pin: Strawberry Shortcake sitting with hearts. $5–10 MOC

3–D plastic pin. $5–10

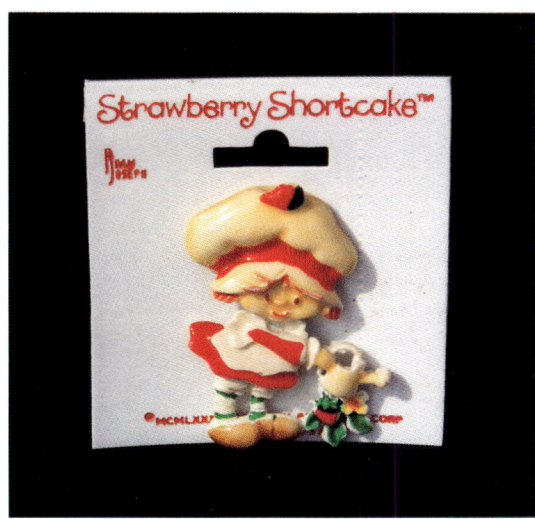

3–D pin with Strawberry Shortcake with watering can. $6–12

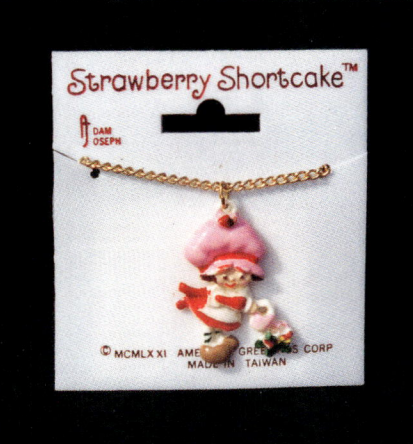

Strawberry Shortcake 3–D necklace. $10–15 MOC

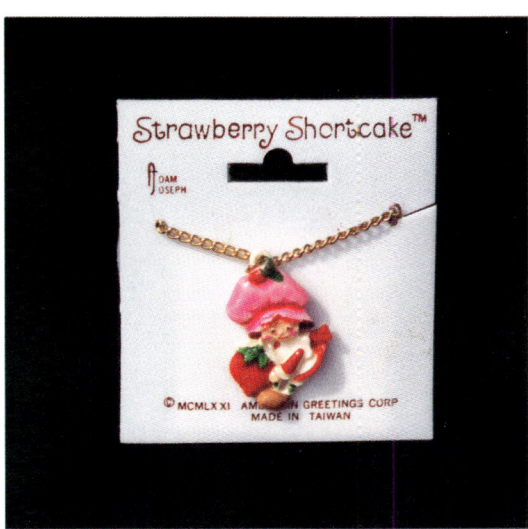

Above: Strawberry Shortcake 3–D necklace, pin and ring. $10–15 each MOC

Left: 3–D ring. $10–15 MOC

The Practical Strawberries

Apple Dumpling Musical. $30–up MIB

Musical Items

Plastic music box. $15–20

Strawberry Shortcake music box. $20–25, Blueberry Muffin $30–35.

Strawberry Shortcake and Cherry Cuddler musical toy, side view.

Back view of musical. $50–65

Plastic battery-operated musical. Berries light up. $25 and up

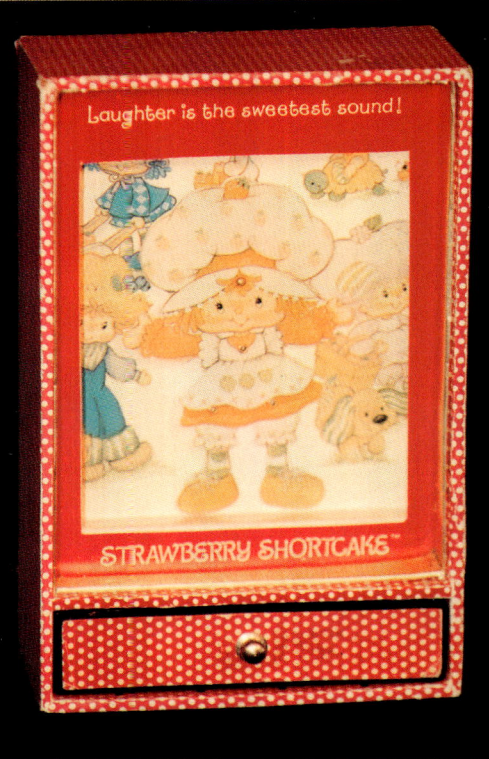

Musical dancing jewelry box. $35–45

Ceramic rotating musical. $30–40

Jewelry box. $30–40

Wooden jewelry box. $45–55

Clocks

Above: Plastic 3–D alarm clock. $30–35 MIB. $15–20 loose

Right: Wall wristwatch clock. $100 and up MIP

Wind-up double-bell alarm clock.
$20–30

Wind-up alarm clock. $35–40

Wind-up double-bell alarm clock with Custard.
$35–50

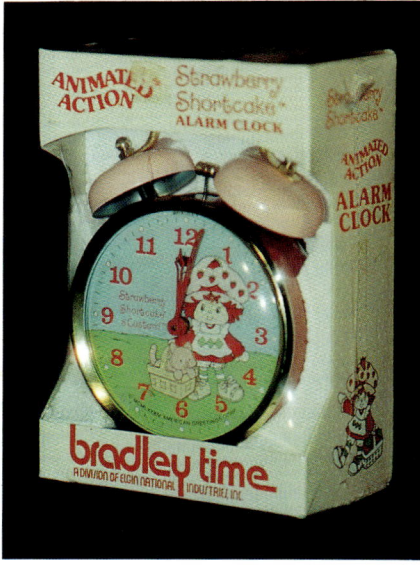

Wind-up double-bell action alarm
clock. $50 and up MIB

Wall clock. $10–15

Wall clock. $15–20

Handmade cloth wall clock. $10–15

Lamp with night light base. $35 and up

Strawberry Shortcake hurricane lamp. Milk glass.
1981. $55 and up

Metal tin-base lamp. $20–25

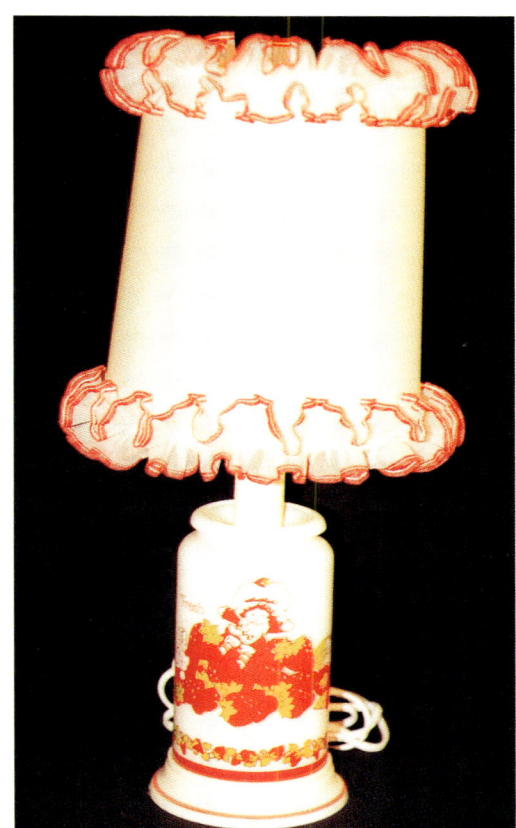

Milk glass lamp with ruffled shade. $25 and up

Porcelain-base lamp: Strawberry Shortcake leaning on strawberry. $30 and up

Metal lamp with strawberries on base.
$30 and up

Motion lamp. $75–up

Strawberry Shortcake lamp. $20–30

Homemade Strawberry shortcake
ceramic lamp. $10–15

Knock-off Strawberry Shortcake lamp. $10–15

*Radios
and More*

Radio. $25–35 MIB

Alarm clock radio. $15–20

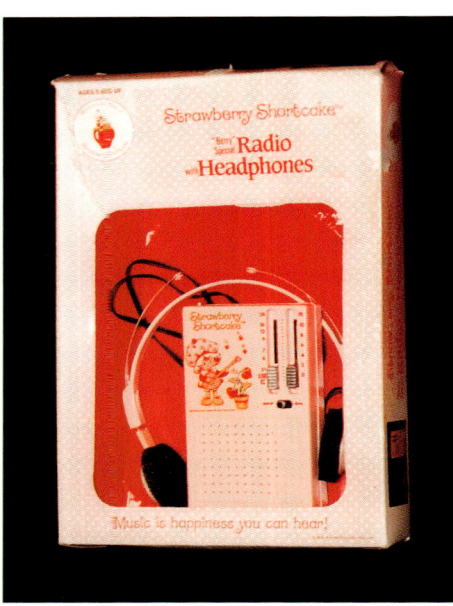

Radio with headphones. $15–20 MIB

Strawberry Shortcake plastic radio. $10–15

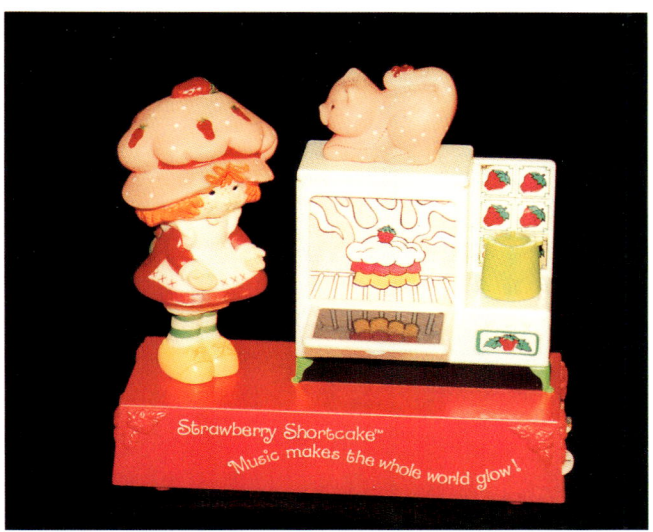

Radio, Strawberry Shortcake holding a cake on a cookie sheet (missing), $15–20. Complete radio, $35–50.

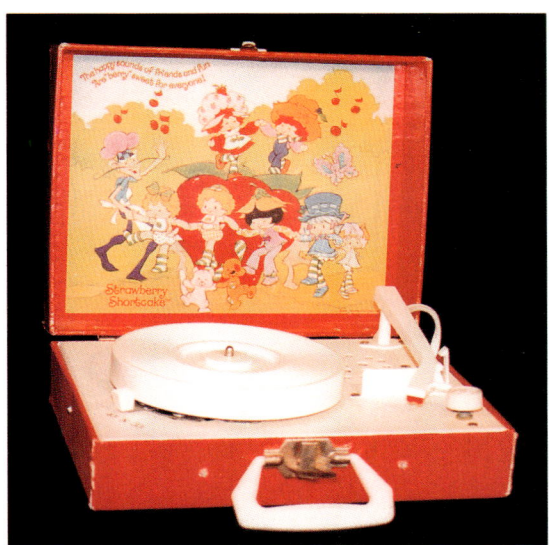

Record player. $35–55

2-dimensional picture, Strawberry Shortcake with Apricot. $25–35

Painted glass picture: Strawberry Shortcake and Lemon Meringue. $15–20

Strawberry Shortcake Mom plaque. $5–9

Strawberry Shortcake wooden plaque. $5–9

Painted glass picture: Raspberry Tart and Strawberry Shortcake. $15–20

Painted glass picture: Strawberry Shortcake. $10–15

Painted glass picture: Strawberry Shortcake and Lemon Meringue under an umbrella. $15–20

Strawberry Shortcake Mom plaque. $5–9

Painted glass picture: Strawberry Shortcake and Lime Chiffon. $15–20

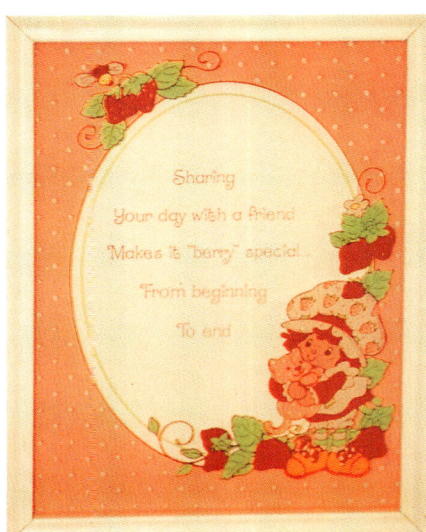

Painted glass picture: Strawberry Shortcake with Custard. $10–15

Painted glass picture: Strawberry Shortcake with Cherry Cuddler. $10–15

Painted glass picture: Strawberry Shortcake with Angel Cake.
$10–15

Strawberry Shortcake and Raspberry Tart
plaque. $5–9

Plaque. $8–12

Wooden and vinyl Strawberry Shortcake plaque. $10–15

Vinyl Strawberry Shortcake plaque. $10–15

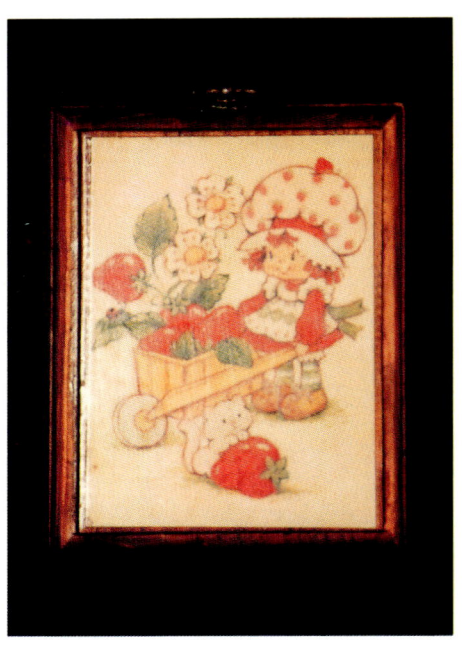

Wooden Plaque. $8–12

Large picture in frame. $20–30

Deco Craft switch plate. $7–10

Plaques. Licensed. $5–10. Homemade $4–6

Handmade cloth stuffed Strawberry Shortcake and Huckleberry Pie picture. $5–10

Plastic 3–D Strawberry Shortcake personalized picture. $10–15

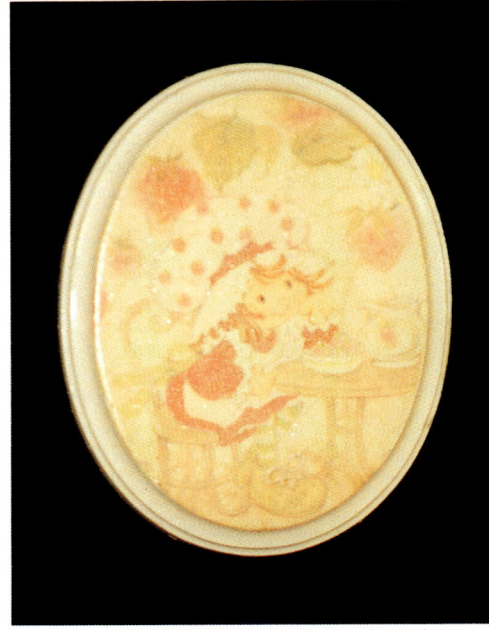

Crystallized picture. $5–10

Crystallized oval picture. $5–10

Framed handmade oval pictures. $5–10

Handmade ceramic wall decoration. $15–20

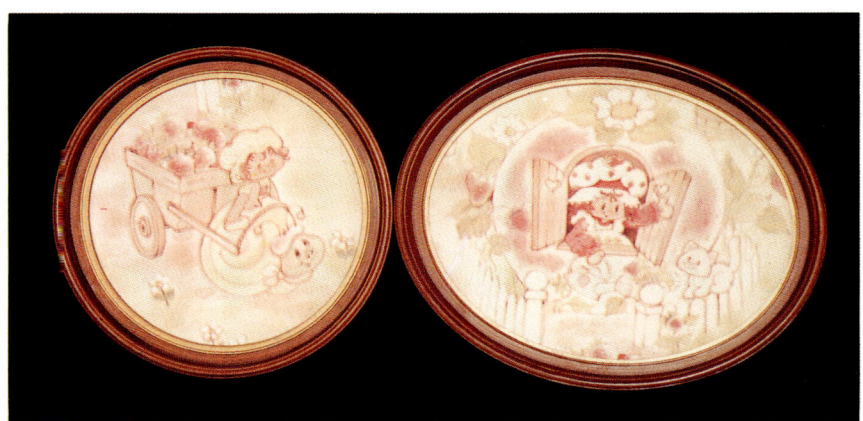

Handmade picture. $5–7

Designer doll house. $75 and up MIB

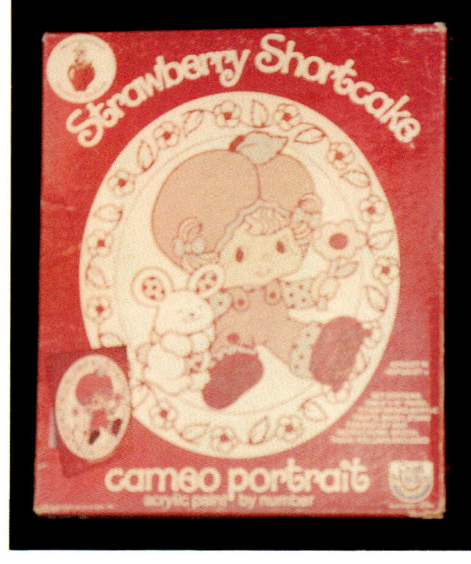

Apricot paint-by-number set. $15–20 MIB

Draw rings by Kenner. $40 and up MIB

Above: Paint by number sets, completed. Lemon Meringue, Strawberry Shortcake, Apple Dumpling. $5–8 each

Below: Paint-by-number sets, completed. Huckleberry Pie, Apricot, Blueberry Muffin. $5–8 each

Craft Master Berry Busy Art center. $40–50

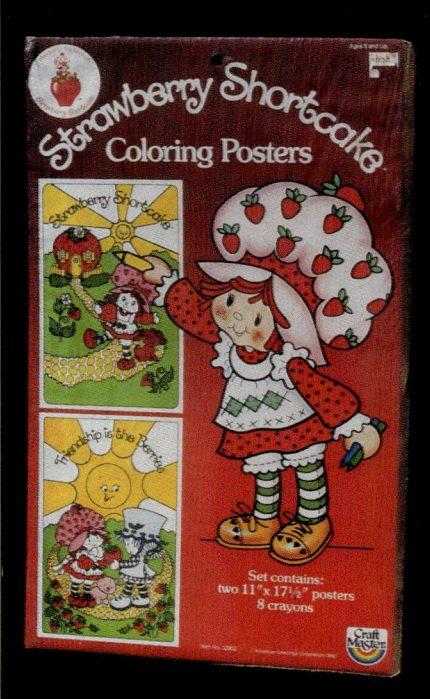

Coloring posters. $20–25 MOC

Color and Sew storybook. $16–22 MIP

Color and Sew story book. $12–18

Coloring postcards. $12–16 MIP

Picture Pretty paint set $16–22

Foreign pictures to color sheet sets. $20–25

Spanish Do and Make Strawberry Shortcake houses for miniatures. 6 different houses. $20–30 each MIP

Playset puzzle. $40 and up

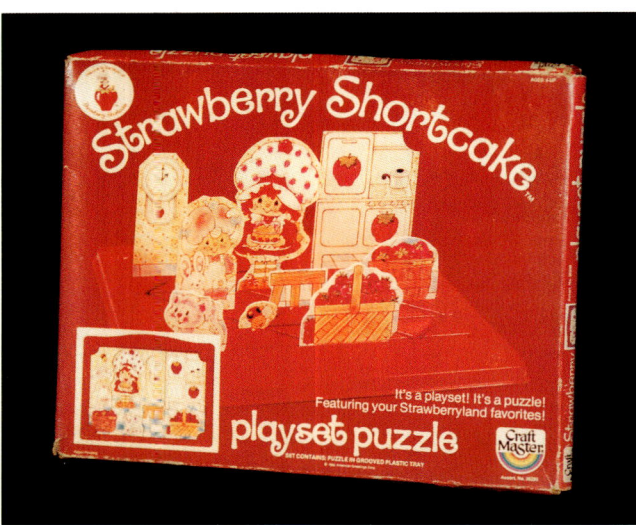

Playset puzzle. 3 sets produced. $40 and up

Make It & Bake It kits. $10–15 each

Latch hook pillow: kit & finished. $20–30

Latch hook rug kit. $16–22

Make It & Bake It jewelry. $10–15 MIB

Make It & Bake It, tabletop set. $5–10 MIP

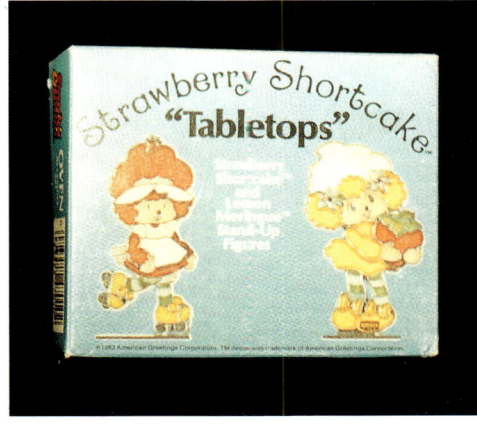

Stain a sticker set. $10–15 MIP

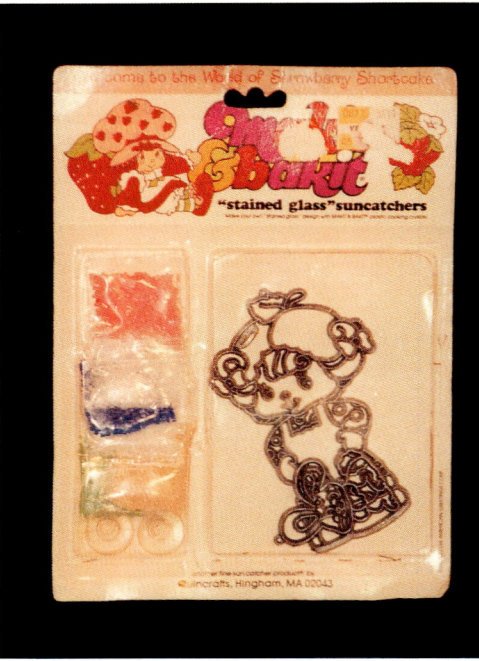

Apricot Make It & Bake It kit. $10–15 MOC

Blueberry Muffin Make It & Bake It kit. $10–15

Make It & Bake It sets, Huckleberry Pie and Cherry Cuddler. $10–15 each MIP

Strawberry Shortcake stained glass Make It & Bake It kit. Mint in box. $25–30

Cherry Cuddler paint a figurine kit. $10–15

Noteboard set. $20–30 MIB

Painted figurines from kits. $5–8 each

Painted figurines. $5–8

Corkboard. $10–15 MIP

Stamp Affair small stamps. $5–6 MIP

Stamp Affair rubber stamps. $5–6 each

Stamp Affair rubber stamps. $5–6 each

Clothing
& Accessories

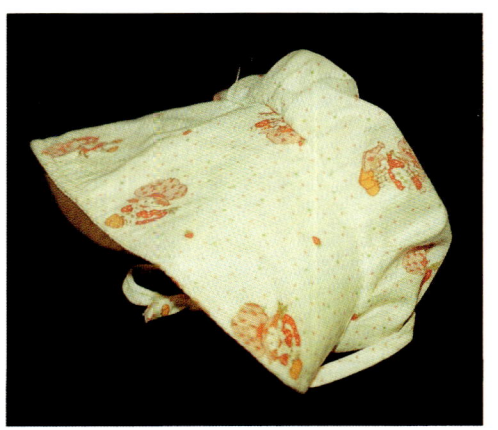

Handmade cotton bonnet. $5–8

Child's scarf. $5–10

Child's cotton dress. $5–8

Child's cotton
shirt. $5–8

Child's cotton
dress. $5–8

Child's dress with
apron. $5–10

Child's plastic apron. Never used. $25–30

Plastic child's apron. Never used. $25–30

Handmade "Friends are the Berries" adult's vest. Price unavailable.

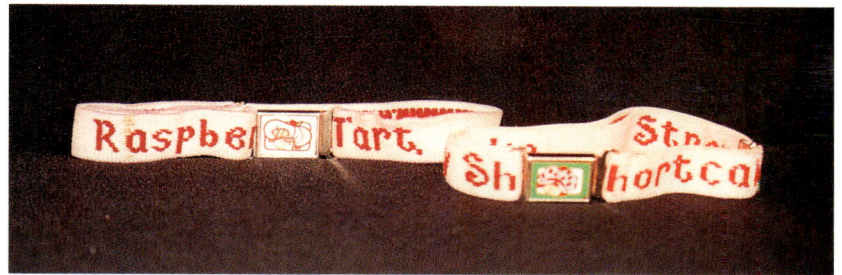

Loose stretch character belts. $5–7

Loose stretch belts. $4–6 each

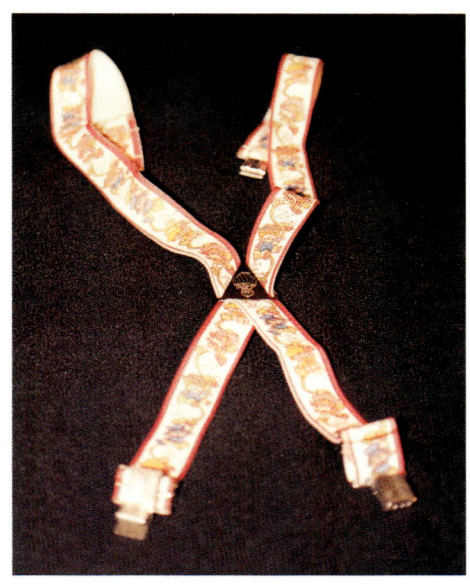

Suspenders. $4–6

Berry belt. $10–15 MIP

Berry belts. Elastic waist with metal buckle. $10–15 MIP each

Berry belt. $10–15 MIP

Berry belt. $10–15 MIP

Ben Cooper
costumes. Lemon
Meringue and
Apple Dumpling.
$10–20 MIB

Ben Cooper cos-
tumes. Sour
Grapes and Purple
Pieman. $10–20
each

Ben Cooper
costumes. Straw-
berry Shortcake
and Apricot. $10–
20 MIB each

Plastic roller skates.
$12–20

Child's roller skates.
$15–20 MIB

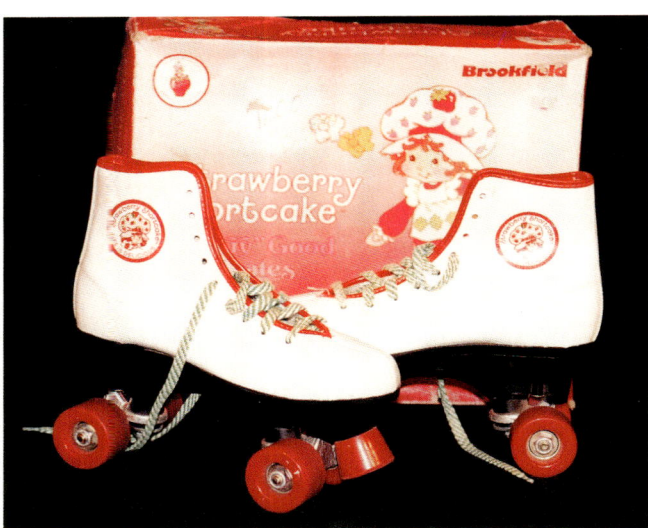

Ice skates.
$15–20 MIB

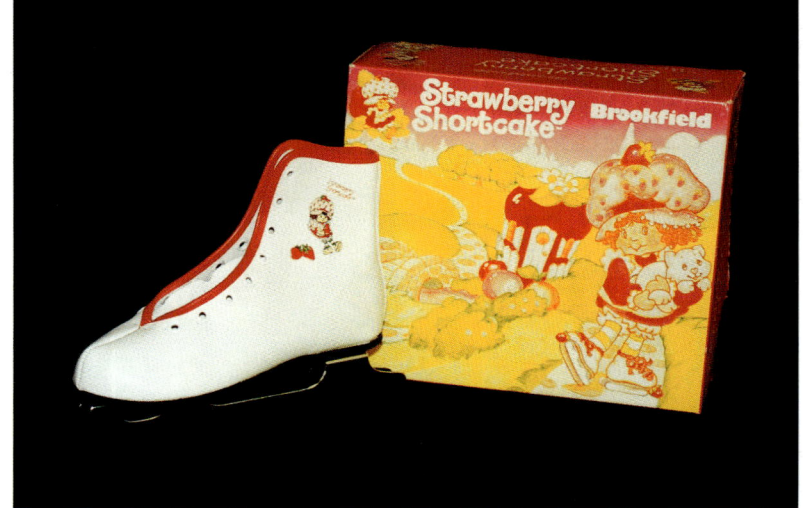

Child's shoes and box. $20–30

Socks. $8–12.

Socks. $10–15 MIP

87

Book bag. $15–20 MIP

Purse with hang tag. $10–15

Purse. $1C–15 MIP

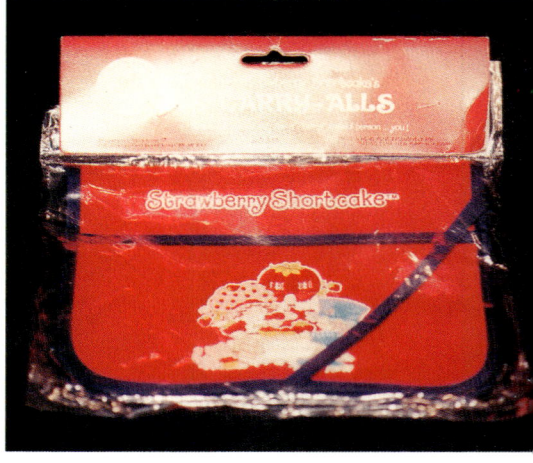

Purse. $4–8 MIP

Purse set. $25–30 MIB

Purse. $10–12 MIP

Spanish purse
"Rosita Fresita".
$8–15

Small canvas purses. $5–10 each

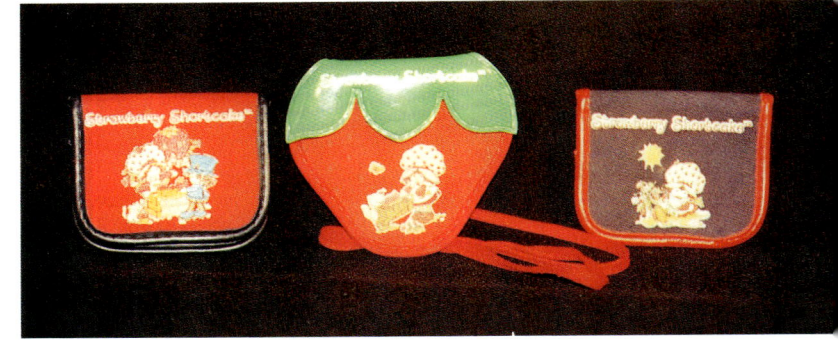

Canvas wallet. $5–8

Large canvas purse. $8–12. Small size, $4–6.

Canvas wallets. $5–10

Butterick pattern. $5–8

Butterick pattern. $3–6

Butterick pattern. $3–6

Butterick pattern to make a doll. $3–6

Butterick pattern. $5–8

Butterick pattern. $5–8

Household Textiles & Accessories

Handmade pillow. $5–10

Home made pillow. $3–5

Sleeping bag. $7–14

Wall hanging. $15–20

Huckleberry Pie wall hanging. $12–18

Table cloth with scenes from Big Apple City video. $30–40

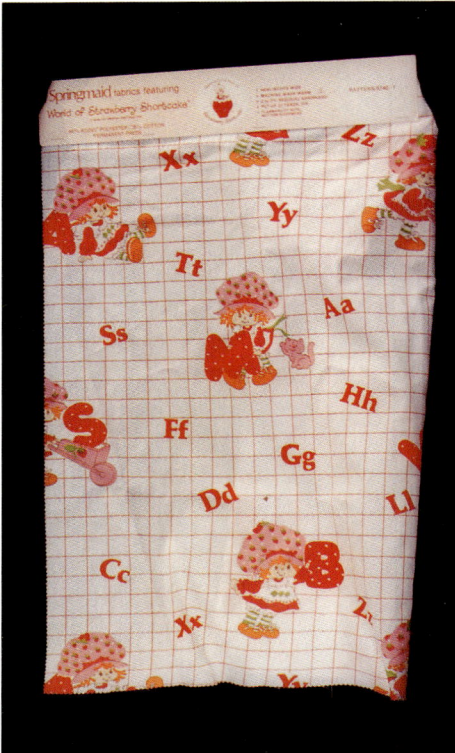

Salesman's sample of Strawberry Shortcake sitting on bows, fabric. Price not available

Salesman's sample of Strawberry Shortcake alphabet fabric. Price not available

Canopy set. $15–20 MIP

Strawberry Shortcake vanity set. $25–up

Apricot toothbrush holder. $50–60 MIB

Strawberry Shortcake toothbrush and cup set. $50–60

Barrettes. $4–8 MIP

Travel set. $10–15

Lemon Meringue vanity set. $25 and up

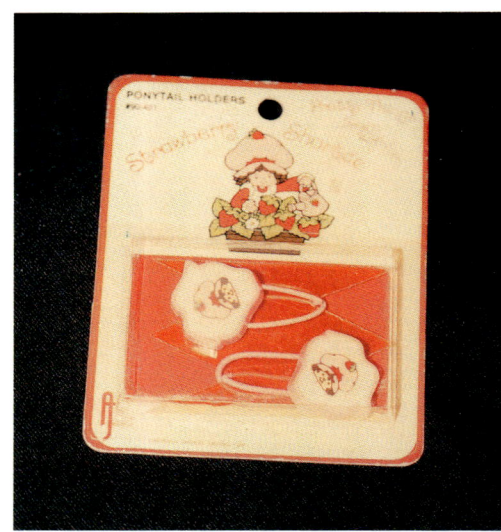

Ponytail holders. $8–10 MIP

Ponytail holders. Cupcake, Strawberry Shortcake, Pupcake. $5–10 each MIP

Barrettes. $5–10 MIP

Sunny Suds Shampoo kit. $20–25

Child's bubble bath and shampoo bottles. $5–10

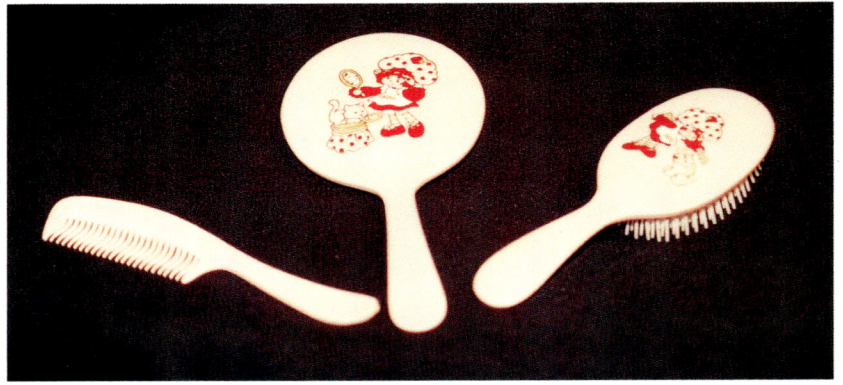

Comb, mirror, brush set. Loose $5 each

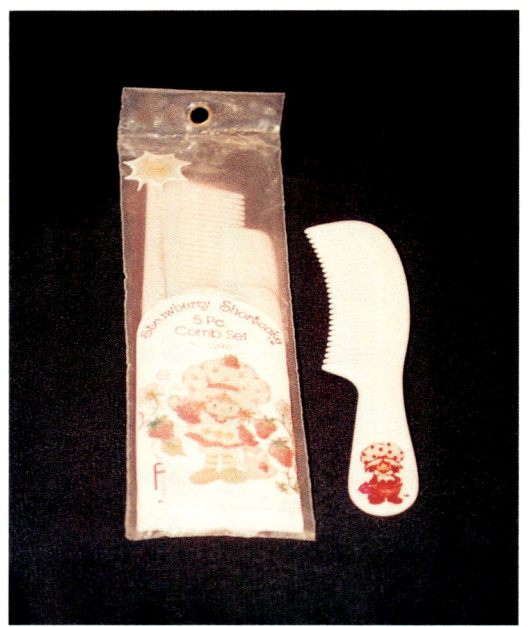

Comb set. $8–14 MIP

Deluxe beauty set. $25–30 MIB

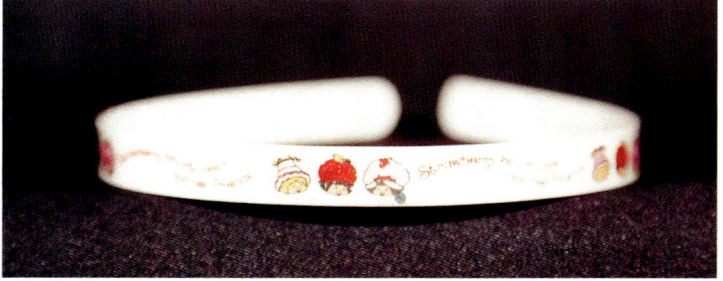

Plastic Strawberry Shortcake child's headband. $10–15

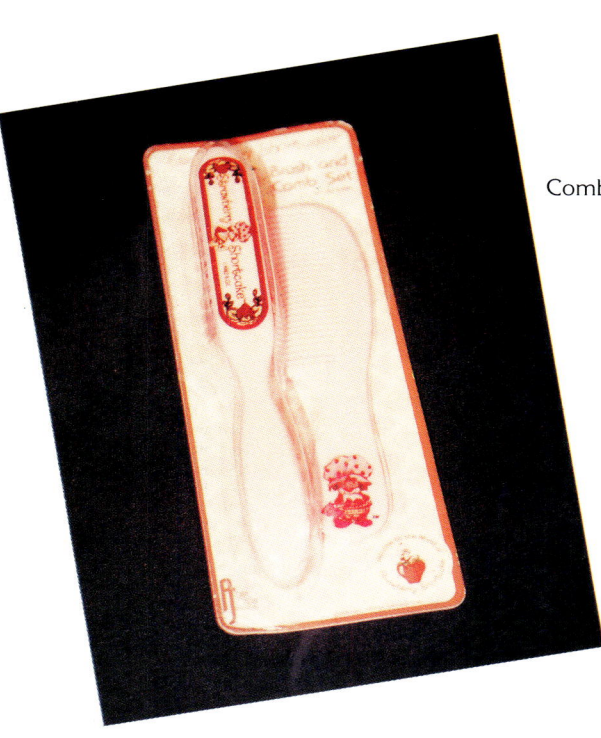

Comb and brush set. $12–20 MIP

Lemon Meringue mirror. $4–6

Blueberry Muffin mirror. $12–18 MIP

"Berry Sweet" mirror. $10–15 MIP

Blueberry Muffin and Strawberry Shortcake mirror. $10–15 MIP

Strawberry Shortcake with Custard mirror. $12–18 MIP

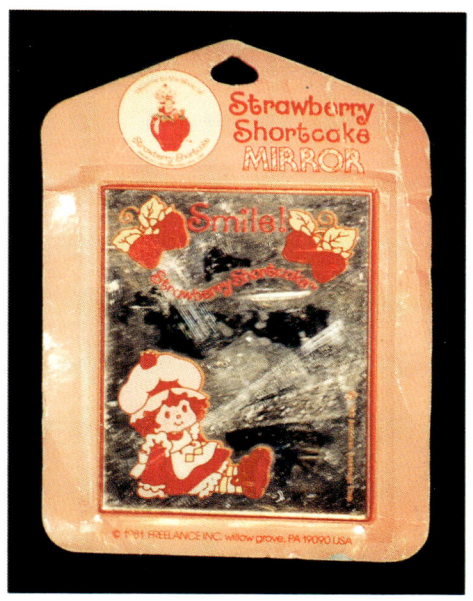

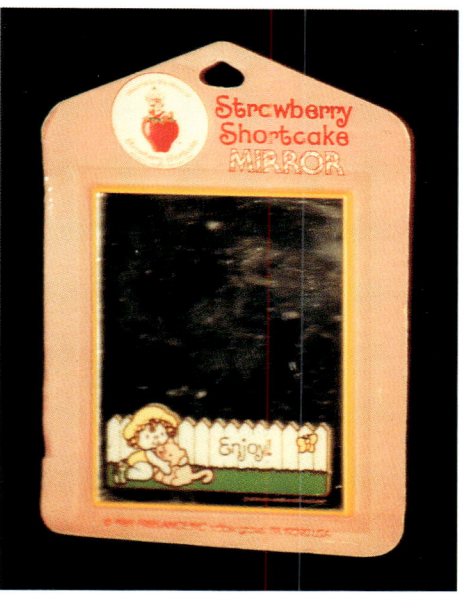

Apple Dumpling mirror. $12–18 MIP

"Smile" mirror. $10–15 MIP

Small mirrors. $5–10 each MIB

Games, Puzzles & Books

Musical Matchup Atari game. $15–20

Strawberryland game. $10–15

Big Apple City game. $10–15

Berries to Market game. $10–15

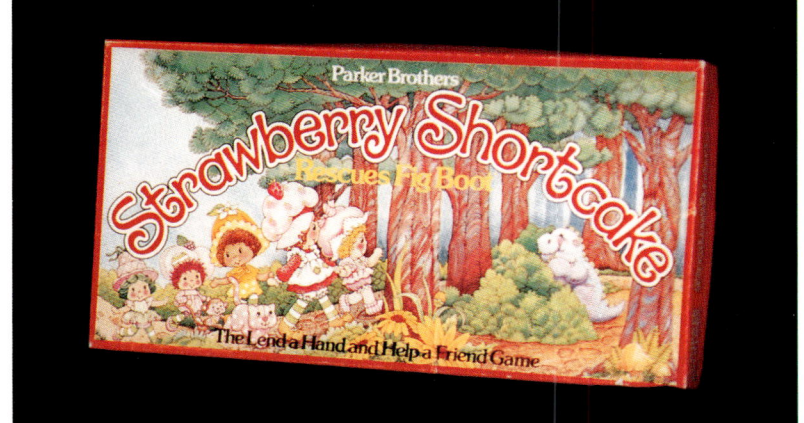

Strawberry Shortcake Rescues Fig Boot game. $10–15

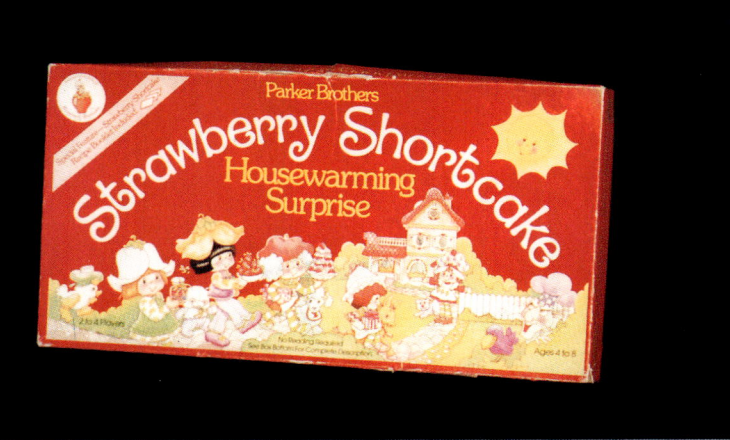

Housewarming Surprise game. $10–15

Card games. $5–10

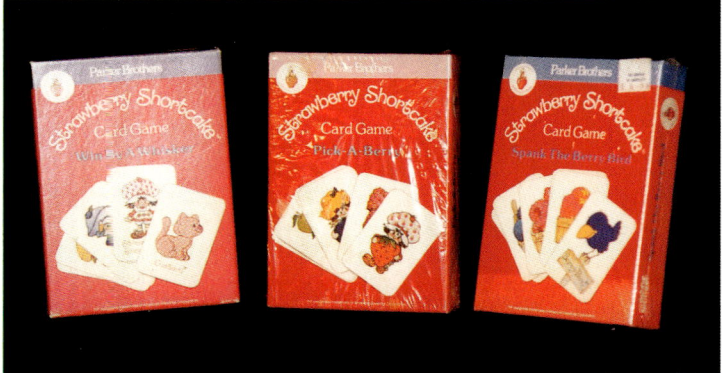

Card games. $5–10

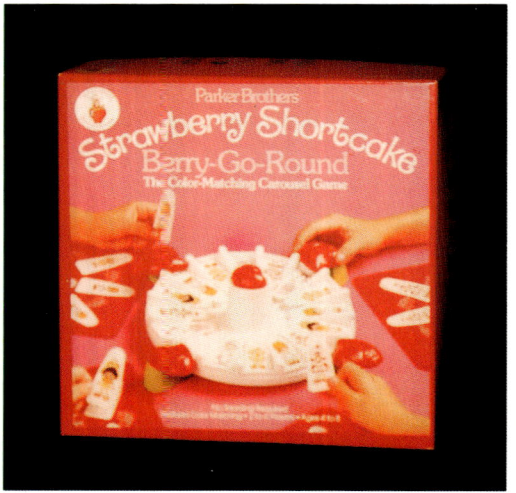

Berry Go Round game. $12–16

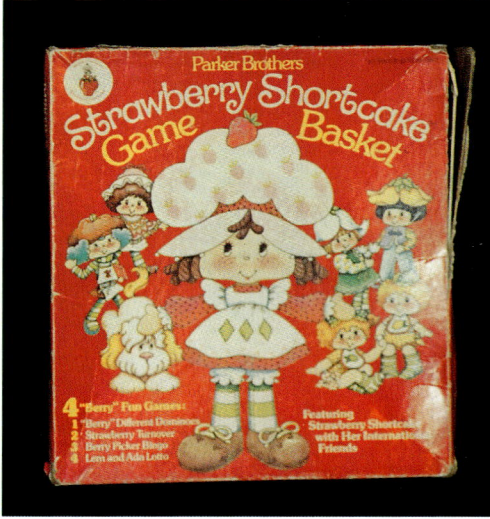

Game basket with International Friends.
$10–15

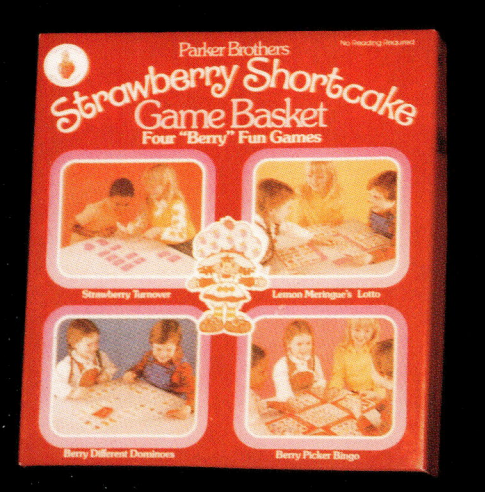

Game Basket. Also produced with the
International Friends version. $10–15

Recipe book. $20–30 MIP

Recipe book. $20–30 MIP

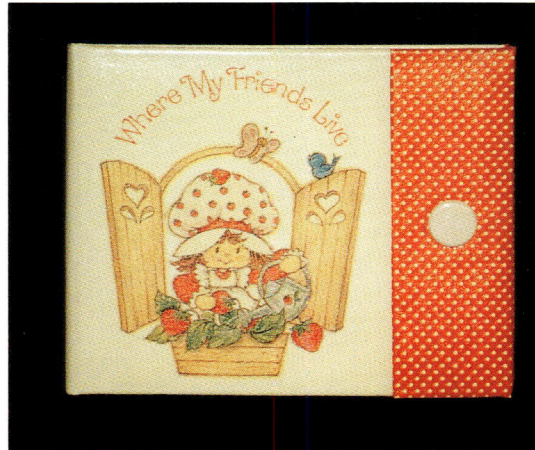

Address book. $15–20

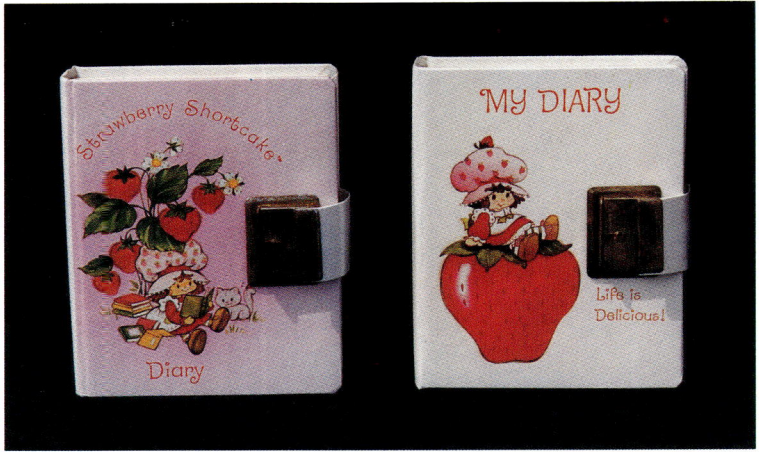

Diaries with keys. $15–20 MIP

Autograph books. $10–15 MIP

Berrykins autograph book. $35–45 MIB

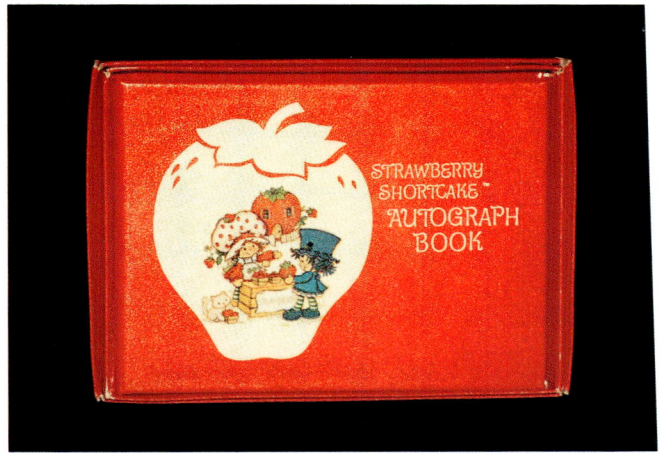

Autograph book. MIB $15–20

Coupon holder and auto record
keeper. $20–25 each

Calendar and coupon holder.
$20–25 each

Small Berrykin puzzle. $5–10

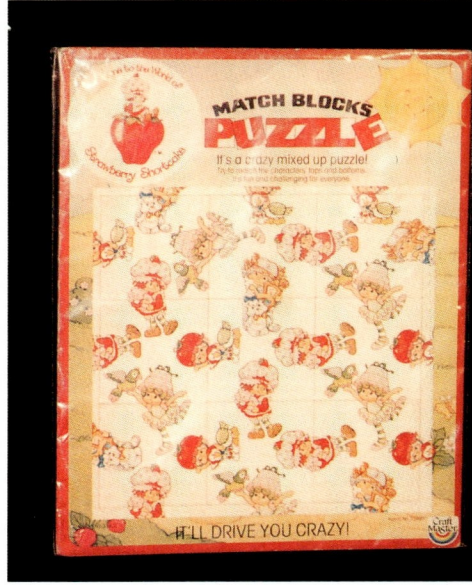

Block puzzle. $9–16

Puzzle. $5–3

Puzzle. $5–8

Puzzle. $5–8

Puzzle. $5–8

Large puzzle, 11" x 14". $6–10

Strawberry Shortcake puzzle.
Craftmaster. $8–10

Large puzzle, 11" x 14". $6–10

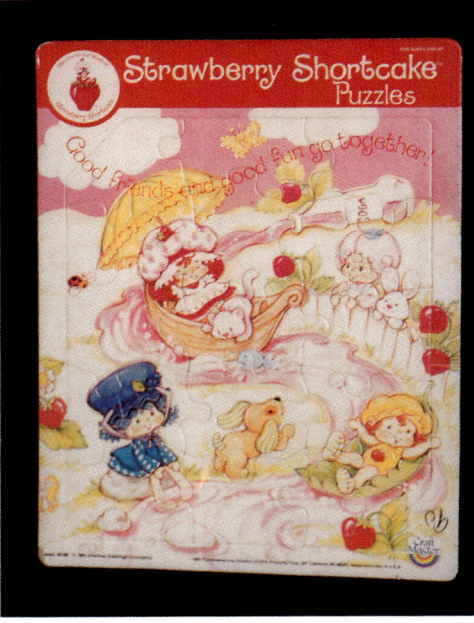

Large puzzle, 11" x 14". $6–10

Large puzzle, 11" x 14". $6–10

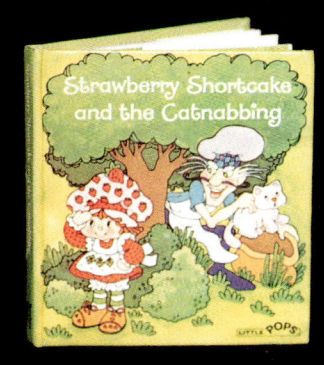
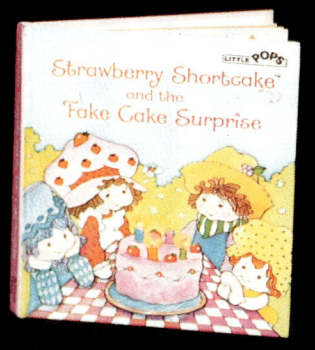

Little Pops books.
"Strawberry Shortcake and the Catnapping" and "Fake cake surprise". $5–10 each

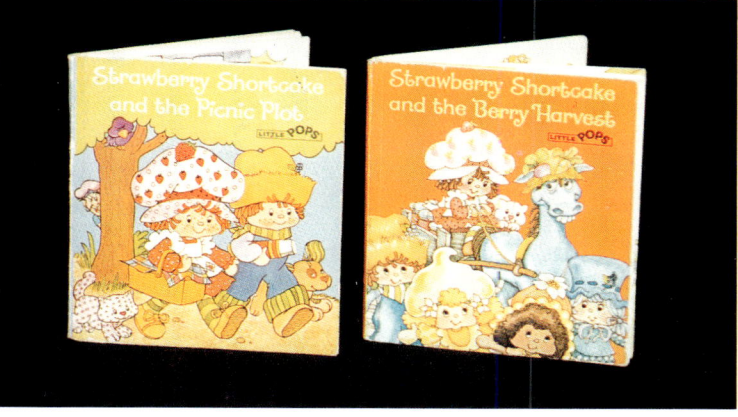

Little Pops books.
"Strawberry Shortcake and the Picnic Plot" and "Berry Harvest." $5–10 each

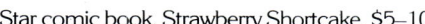
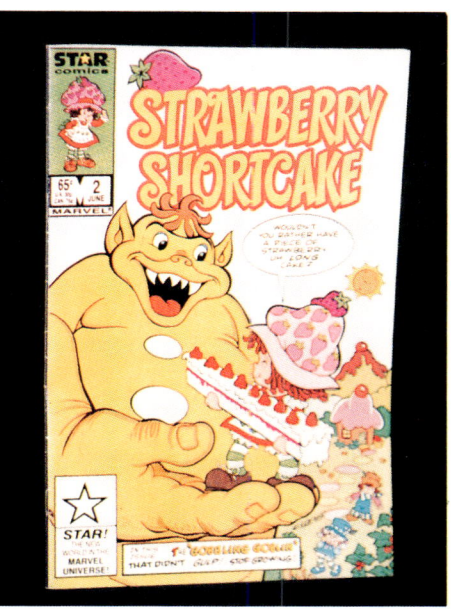

Star comic book, Strawberry Shortcake. $5–10

Strawberry Shortcake Star comic. $10–20

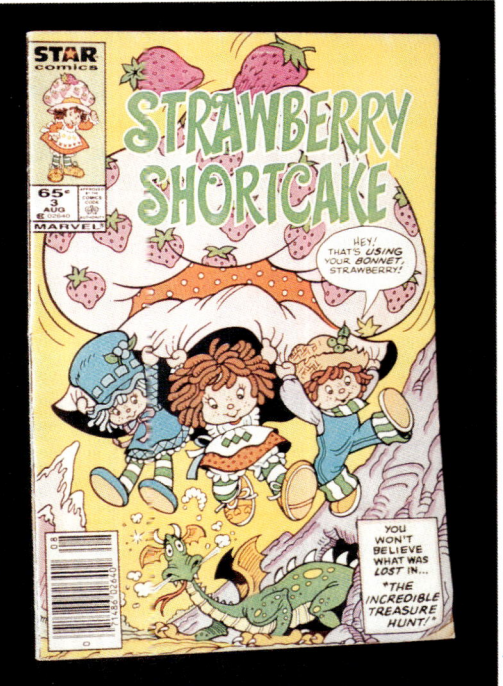

Star comic book. $10–20

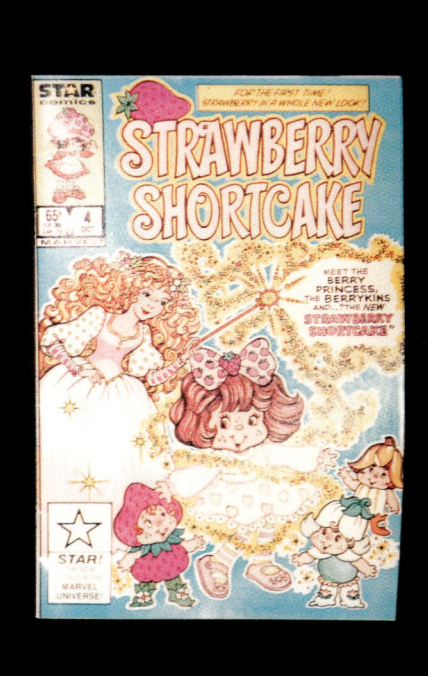

Strawberry Shortcake comic book. $10–20

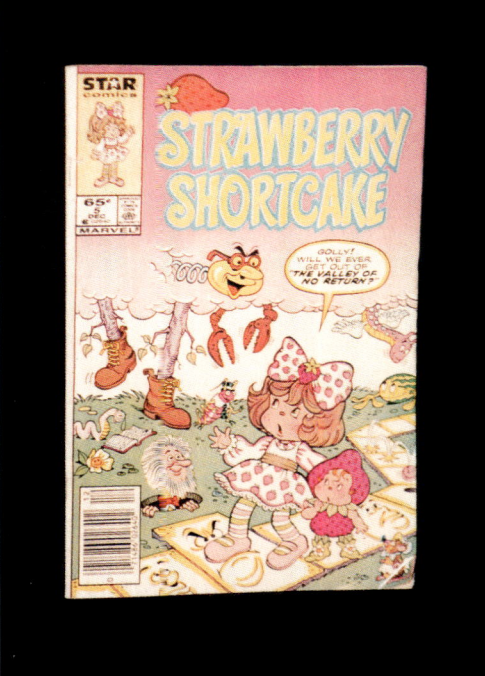

Strawberry Shortcake comic book. $10–20

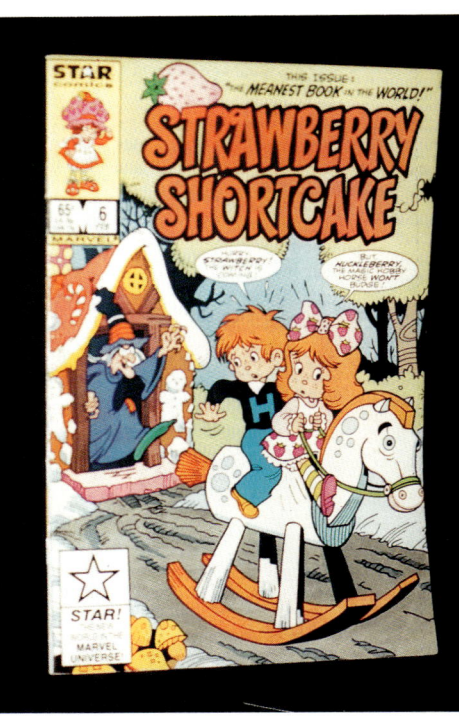

Strawberry Shortcake comic book. $10–20

Paragon needlepoint book. 1985. Plum Pudding, Peach Blush etc. $20 and up

Strawberry Shortcake and Berry Bears. $5–8

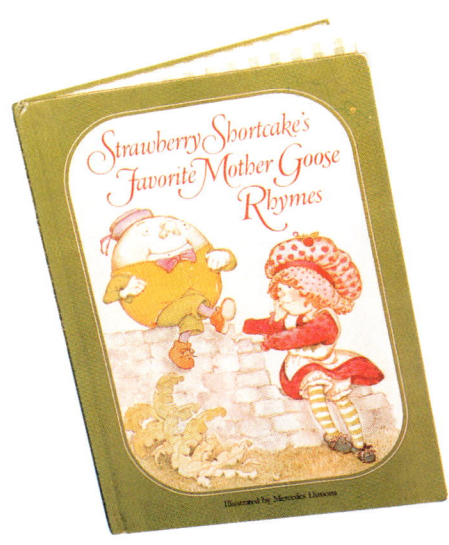

Hard cover book, *Strawberry Shortcake's Favorite Mother Goose Rhymes*. Parker Bros. $6–8

Hard cover book, *Strawberry Shortcake and the Winter that Would Not End*. $4–6

Holiday library set with four books and holder. $10–15

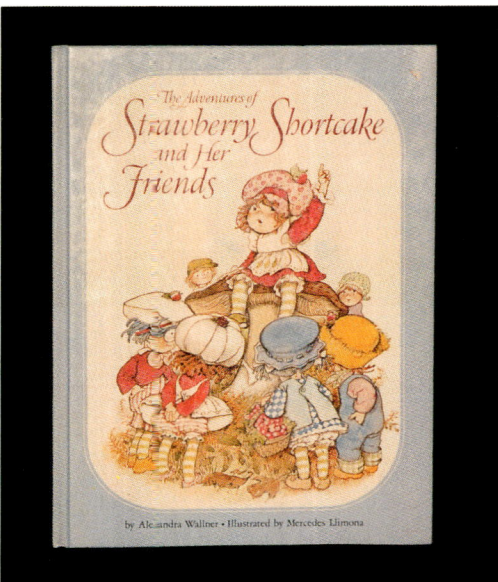

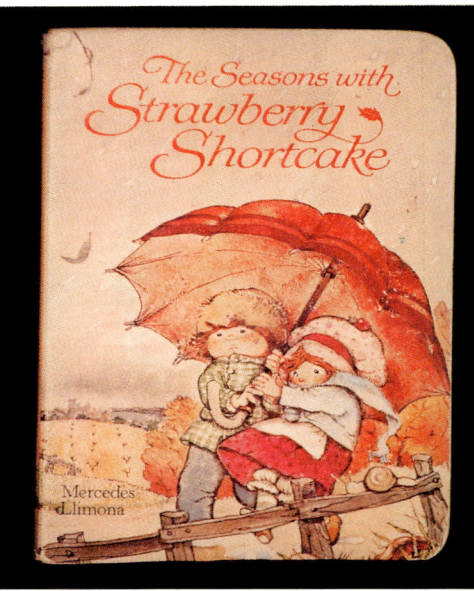

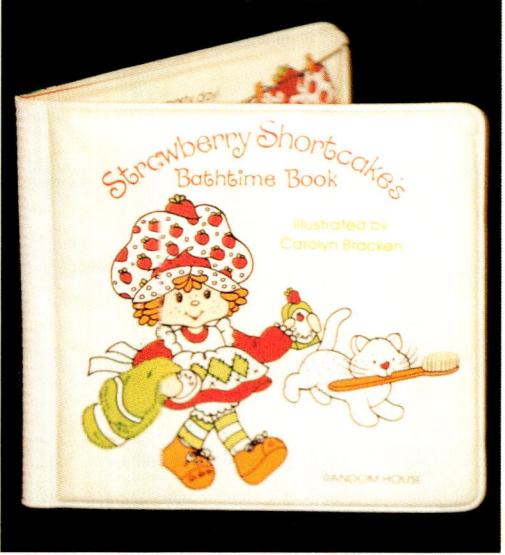

Above Left: Hard cover book, *Strawberry Shortcake and her Friends*. $5–10

Above: Seasons with Strawberry Shortcake book. $5–10

Left: Vinyl bath book. $4–6

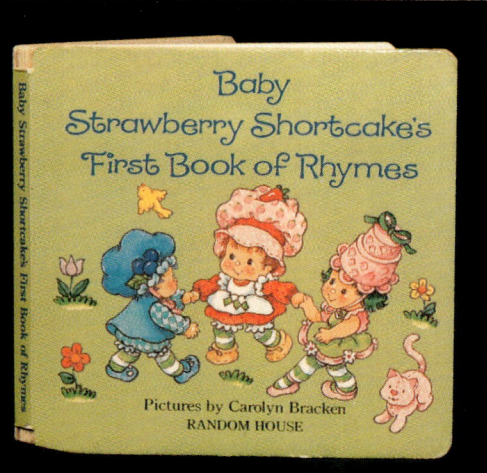

Baby Strawberry Shortcake's First Book of Rhymes. Random House. $6–10

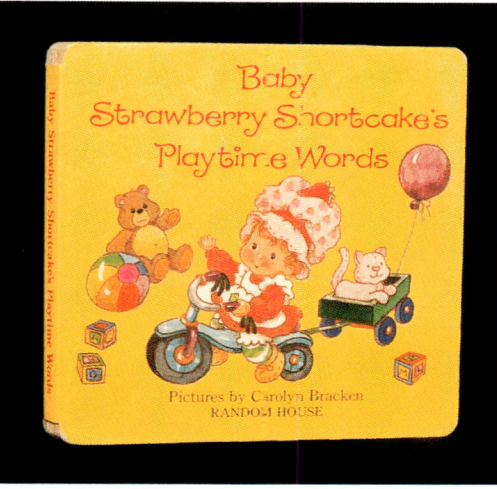

Baby Strawberry Shortcake's Playtime Words. Random House. $5–10

Preschool book. $5–10

Party fun book. $4–8

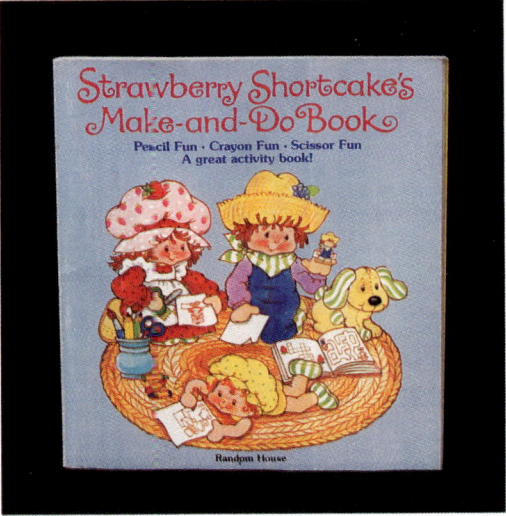

Make and Do activity book. $5–10

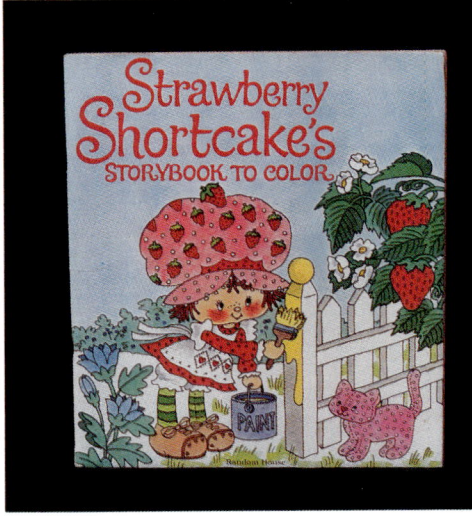

Storybook to color. $5–10

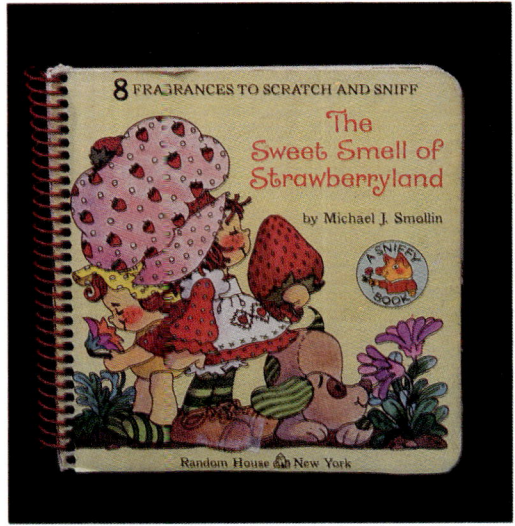

Sweet Smell of Strawberryland scratch and sniff book. $5–10

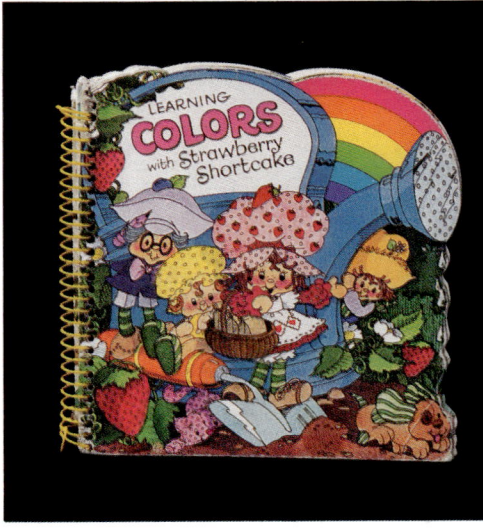

Learning Colors book. $5–10

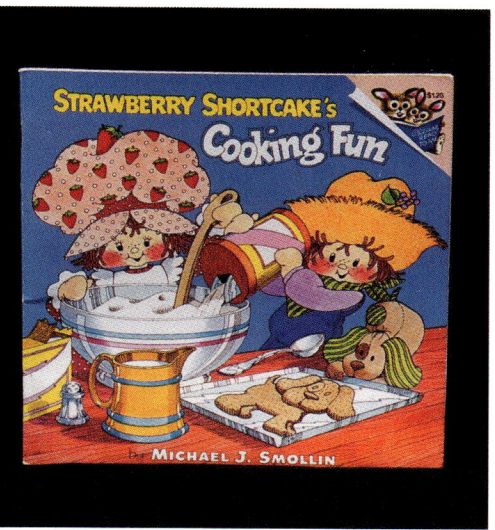

Cooking Fun book. $4–8

Big Apple City activity book. $8–12

Strawberry Shortcake coloring book.
Kenner. $6–8

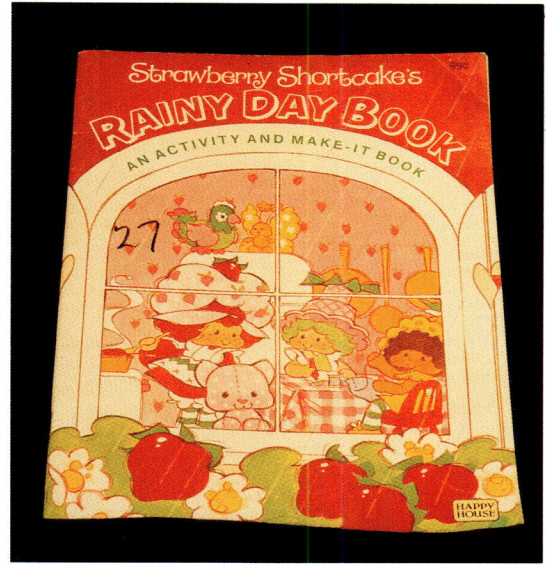

Rainy Day activity book. Happy House. $4–6

Coloring book with Strawberry Shortcake
and Custard. $5–10

Book of Mazes. $5–10

Busy book. $5–10

Birthday party coloring book. $8 and up Mint

Christmas coloring book. $5–8

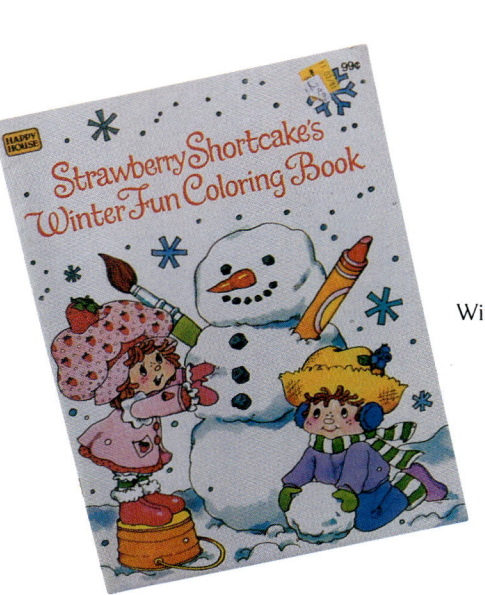

Winter Fun coloring book. $8–12

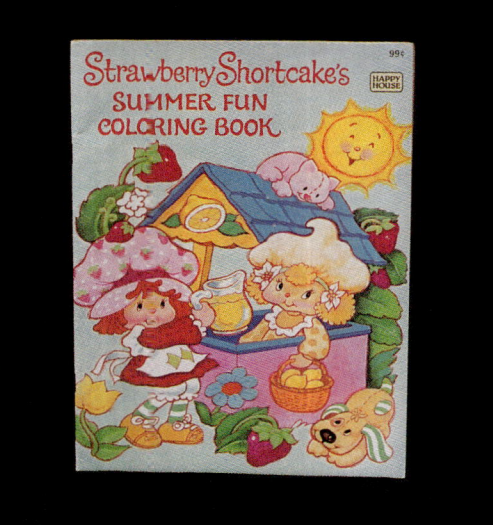

Summer Fun coloring book. $10–15

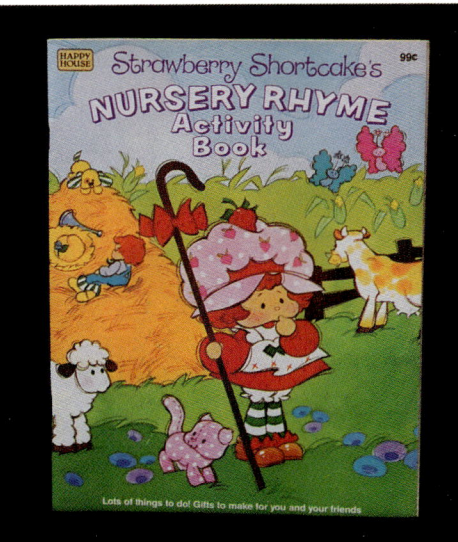

Nursery rhyme coloring book. $10–15

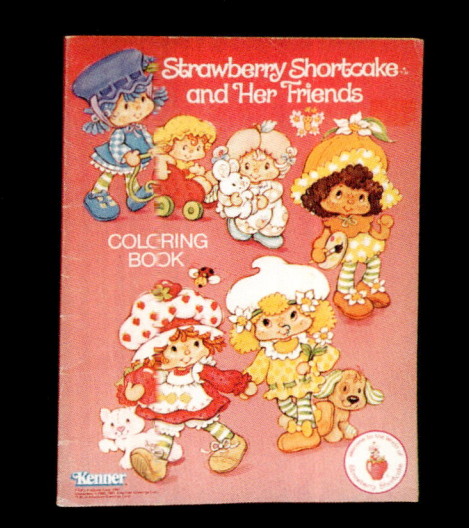

Coloring book. $5–10

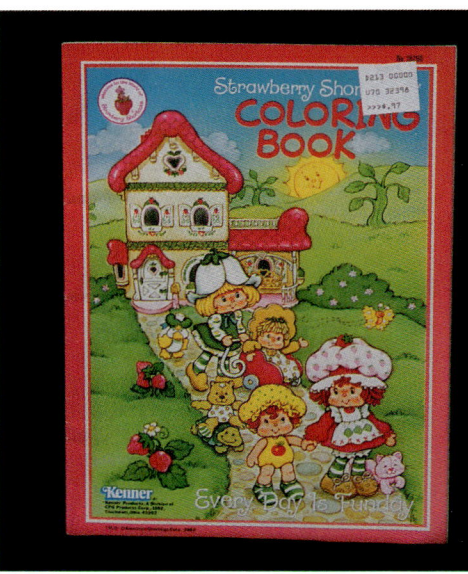

Coloring book. $5–10

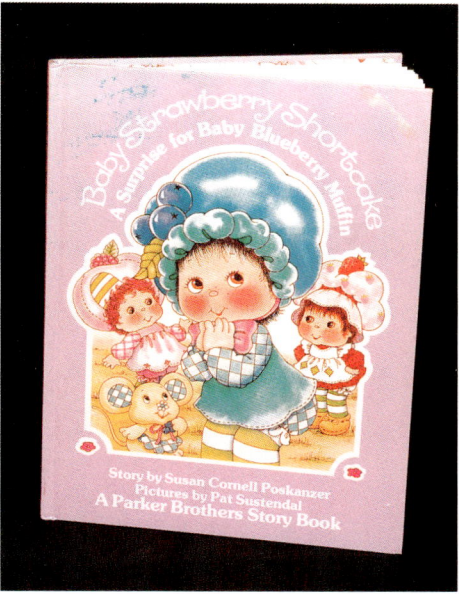

Baby Strawberry Shortcake. *A Surprise for Baby Blueberry Muffin*. Parker Bros. $6–8

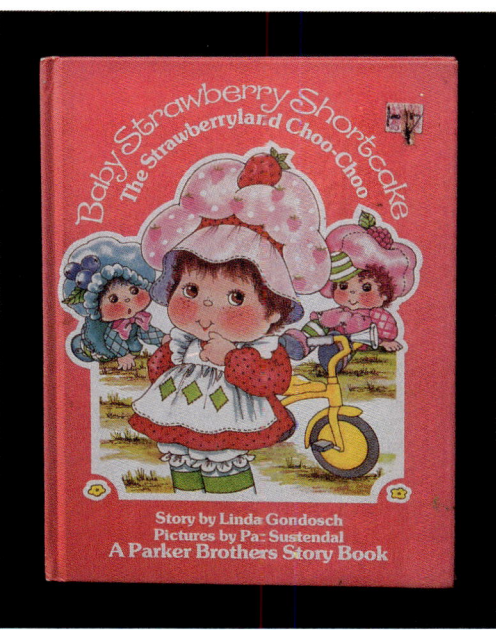

Parker Bros. Strawberryland Chco Choo. $5–10

Baby Strawberry Shortcake. *Fig Boot's Happy Day.* Parker Bros. $5–7

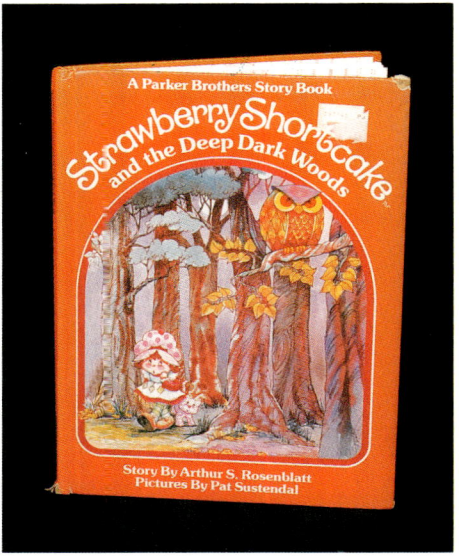

Hard cover book, *Strawberry Shortcake and the Deep Dark Woods*. Parker Bros. $6–8

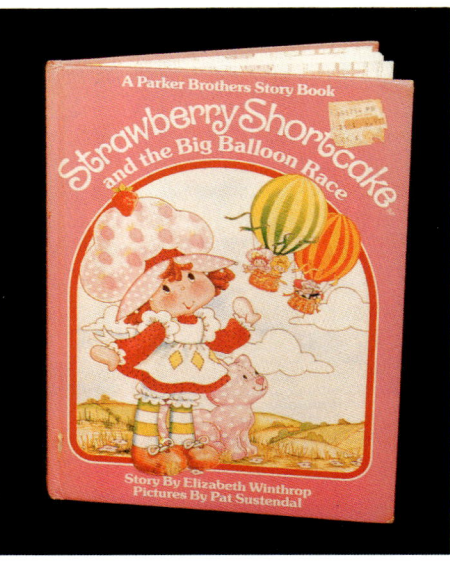

Hard cover book, *Strawberry Shortcake and the Big Balloon Race*. Parker Bros. $6–8

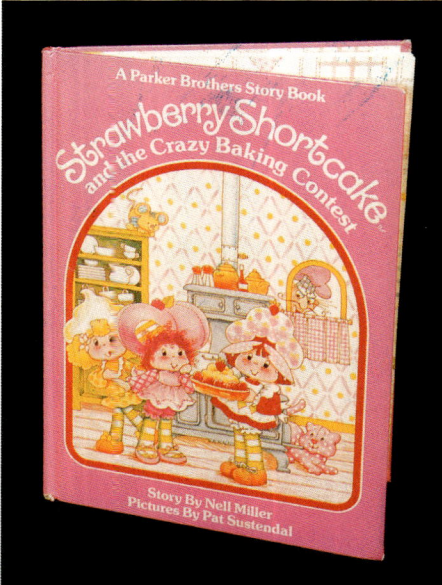

Hard cover book, *Strawberry Shortcake and the Crazy Baking Contest*. Parker Bros. $8–10

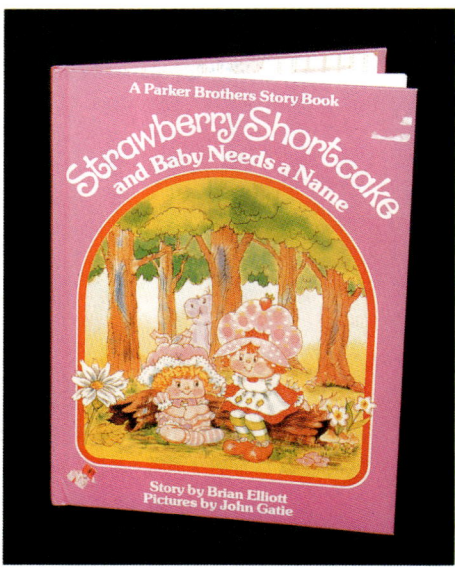

Hard cover book, *Strawberry Shortcake and Baby Needs a Name*. Parker Bros. $8–10

Hard cover book, *Strawberry Shortcake and the Birthday Surprise*. Parker Bros. $8–10

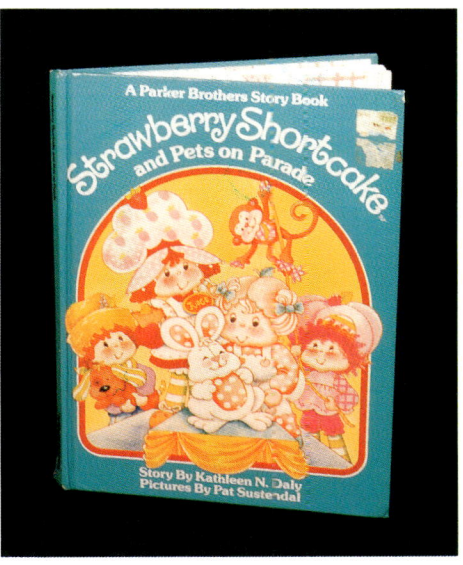

Hard cover book, *Strawberry Shortcake and Pets on Parade*. Parker Bros. $6–8

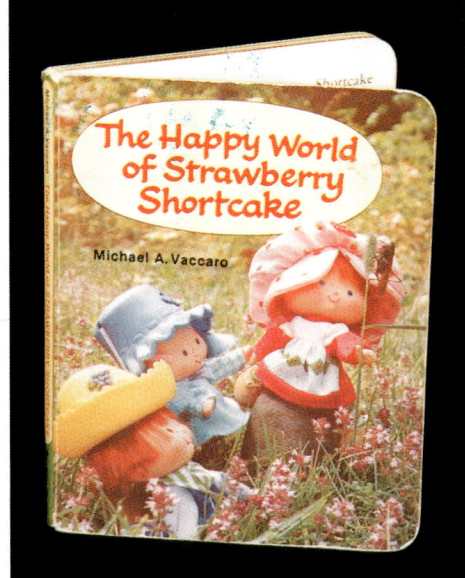

Hard cover book, *The Happy World of Strawberry Shortcake*. Michael Vaccaro. $7–10

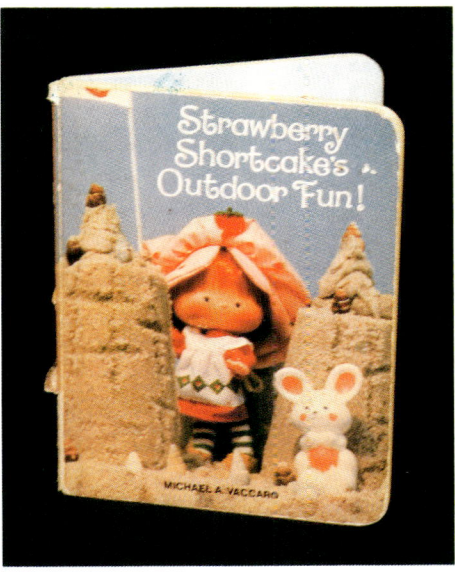

Hard cover book, *Strawberry Shortcake's Outdoor Fun*. Michael Vaccaro. $7–10

Kitchen

Beverage glasses. $10–15 MIB

Small beverage glasses.
$10–15 MIB

Beverage glasses. $10–15 MIB

Salt and pepper shakers.
$10–15 MIB

3-piece flowered glass
canister set. $25–35 set

Flowered glass juice
carafe, $15–20. Large
and small matching
glasses. $3–5 each

Juice carafe with 4 small glasses. $15–20 MIB

Child's silverware set. $45 and up

Milk glass Raspberry Tart and Strawberry Shortcake mugs. $4–6 each.

 Milk glass Huckleberry Pie and Apple Dumpling mugs. $4–6 each

White milk glass mugs. Strawberry Shortcake with watering can, Apricot, Orange Blossom. $20–30 each

Milk glass bowl. $4–6

Plastic pitcher with lid. $15–20

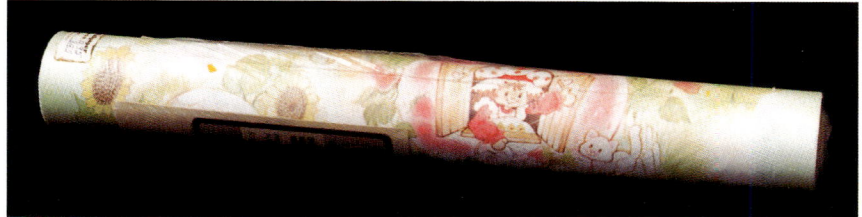

Roll of wallpaper, still in package. $20 and up

Plastic child's character bowl and cup. $10–15

Plastic child's cups. 2 sizes. $3–6

Advertisement from magazine for Betty Crocker dish set in upper right corner. Lower left corner is the matching dinnerware. $2–3

Plastic sectioned child's pate. $4–6

Betty Crocker special mail-away set. $20–30

Plastic child's dishes. $20–25 MIB

4" x 6" plastic tray of Strawberry Shortcake with polka dots. $20 and up

4" x 6" plastic tray of Strawberry Shortcake and Custard. $20 and up

Large plastic placemats. $10–25

Large placemat. $10–15

Paper towel dispenser. $20–30

Wooden and ceramic tile napkin holders. $25–30

Above: Wooden spice rack. $45–55 MIB

Right: Tissue box cover. $16–22

Handmade apron. $4–8

STRAWBERRY SURPRISE

Wash, hull and slice 1 quart ripe strawberries. Add 2 cups sugar and allow mixture to stand 1 hour. Heat gently over low heat and let berries come to a boil for 1 minute. Remove from heat and allow to cool. Serve over vanilla ice cream with whipped cream.

Strawberry Shortcake adult apron with recipe. $20–25

Cotton handkerchief. $15–20

Christmas towel and oven mitt. $12–17 each

Christmas towel and pot holder. $12–17 each

Pot holder and oven mitt. $6–12

Above: Wash cloth. $4–9

Left: Hand towel, $10–15 mint.

Polka dot lunch bags. $2–4

Lunch bags. $15–20 MIP

Round tin. $5–10

Octagonal storage tin. $15–20

3-piece cookie tin set.
$25–35 set

Small round
storage tins.
$8–12

3-piece tin storage set.
$30–40 set

Round tin with polka dots.
$5–10. Round tin with
different characters. $15–
20. Tin depicting Straw-
berry Shortcake on a pile
of strawberries. $5–10

Small tins of Strawberry Shortcake and friends. $8–12 each

Square tin, $10–15.
Round metal tin or candle holder. $10–15

Above: Metal tin candle holder. $8–15

Left: Metal tin. $5–10

Plastic lunchbox and thermos. $10–15

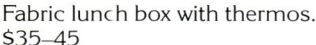
Fabric lunch box with thermos. $35–45

Metal Berrykin lunch box and thermos with hangtags. $40–50

Metal lunch box with thermos.
$10–15

Vinyl lunch box with thermos.
$30–35

Metal lunch box
with thermos.
$10–15

139

Various Strawberry Shortcake thermoses. Berrykin thermos $15–20. Others $5.

Small picnic tin. $15–20

Large picnic tin. $25–30

Lunch tin. $10–15

Metal waste can. $10–15

Metal waste can of Strawberry Shortcake
and Lime Chiffon. $15–20

Metal TV tray of Strawberry Shortcake at picnic. $15–20

Metal waste can of Strawberry Shortcake on
pile of strawberries. $15–20

Metal waste can of Strawberry Shortcake
and Apricot. $15–20

Metal waste can of Strawberry Shortcake
sitting on leaf. $10–15

Metal waste can of Strawberry Shortcake
and Lemon Meringue. $15–20

Metal tray, Strawberry Shortcake with Apricot. $10–15

Metal tray. $10–15

Strawberry Shortcake with Lime Chiffon metal tray. $10–15

Strawberry Shortcake sitting on a pile of strawberries, metal tray. $10–15

Strawberry Shortcake with Lemon Meringue,
metal tray. $10–15

Magnet. $5–8 MIP

Above: Ceramic Strawberry Shortcake
magnets. $5–7

Right: Magnets. $5–10 MIP

Strawberry Shortcake rocking chair. 2 styles. $35 and up

Child's bicycle. $50–60

Child's growth chart and clothes rack. $35 and up

Child's growth chart and clothes rack. $35 and up

Wall hutch. $35 and up

Vinyl doll buggy. $25–35

Imitation Strawberry Shortcake table with 2 chairs. $20–30

Above: Child's chair and step stool. $25–40

Right: Child's easel. $45 and up

Child's wall clothes rack. Loose, $10–up. $20–up MIB.

Above: Coleco plastic doll buggy. $25–35

Left: Doll high chair. $50 and up MIB. $20 and up loose

Vinyl doll buggy. $25–35

Child's Berry Buggy. $50–60 MIB.
$25–30 Loose

Child's telephone set. $10–15

Luggage tags. $4–10 MIP

Child's suitcases. $10–20 each

Child's carry-all bag. $15–20

150

Paper Products

School folders. $5–7 each

School folder. $4–6

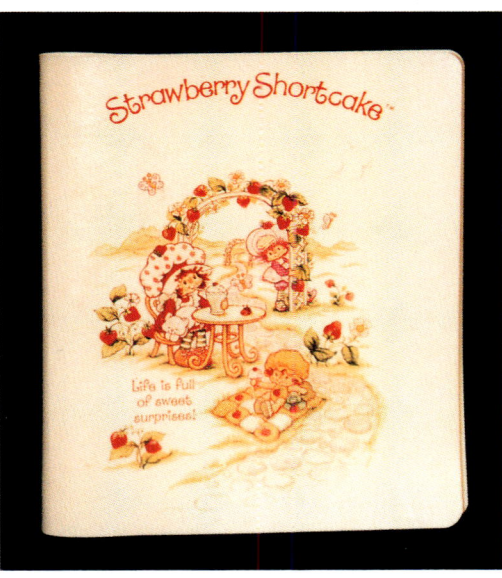

Plastic 3-ring binder. $10–15

School folders. $5–7 each

Notebook cividers. $5–10 MIP

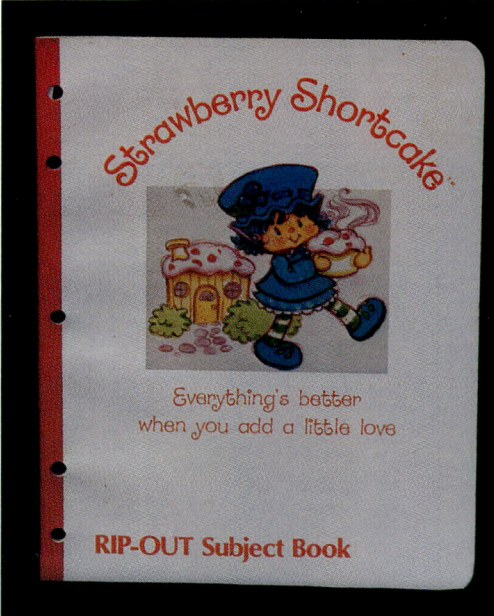

Blueberry Muffin notebook. $5–10

Plastic binder. $10–15

3-ring binder. $10–15

Clip-a-longs. $5–10 MIP

Strawberry Shortcake subject book. $5–7

Apple Dumpling memo book. $5–7

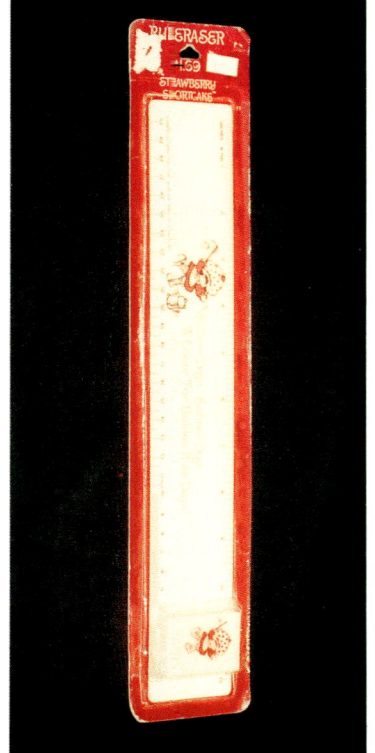

Small memo books. $7–12 each

Ruler in package. $6–12

Picture tape. $8–12

Large clip. $5–7

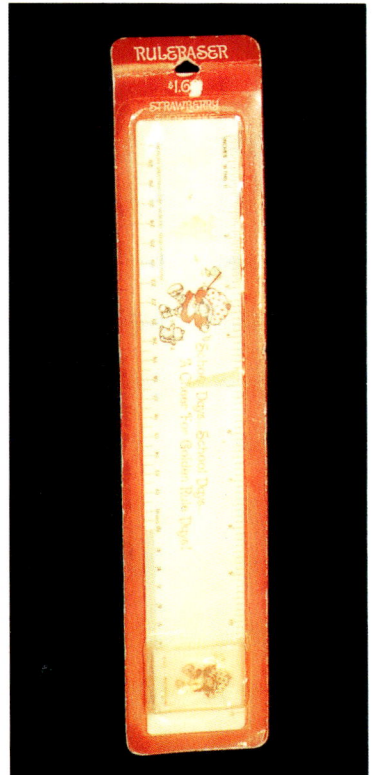

Ruler with eraser in package. $7–12

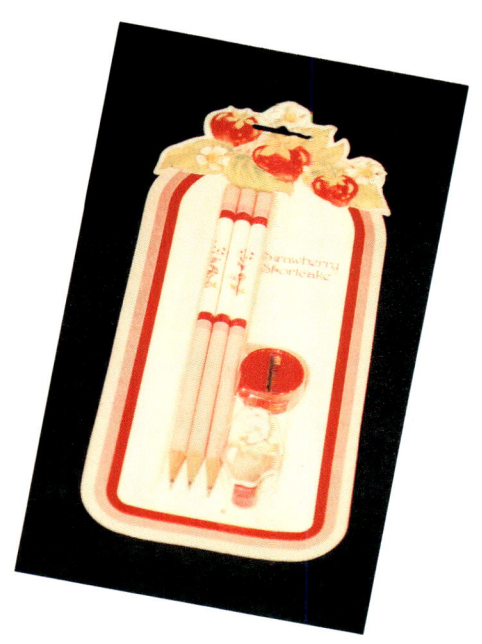

School supplies. Pencils and sharpener. $12–18

155

School supplies. $15–20 MIP

Imitation erasers. $2–4 each MIP

Plastic 2–dimensional container. $10–15

Plastic 2–dimensional container. $10–15

Plastic pencil and notepad holder. $8–12

Markers. $10–15

Pencil holder bag. $4–6 loose

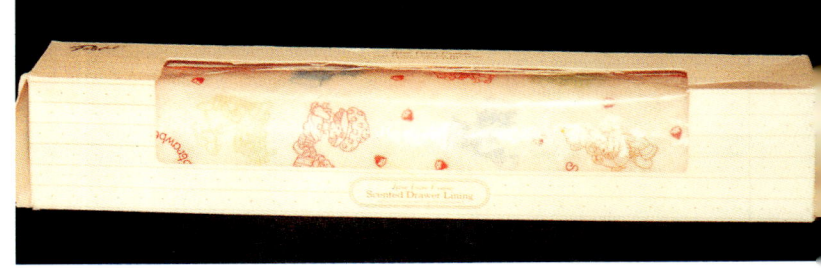

Scented drawer lining. $15–20 MIB

Gift box put together. $4–6

Gift box. $5-10 MIB

Gift box. $5–10 MIB

Small gift box. $4–6 MIP

Gift box put together. $4–6

Gift box. $4–6

Gift box, Strawberry Shortcake with head through wreath. $5–10 MIP

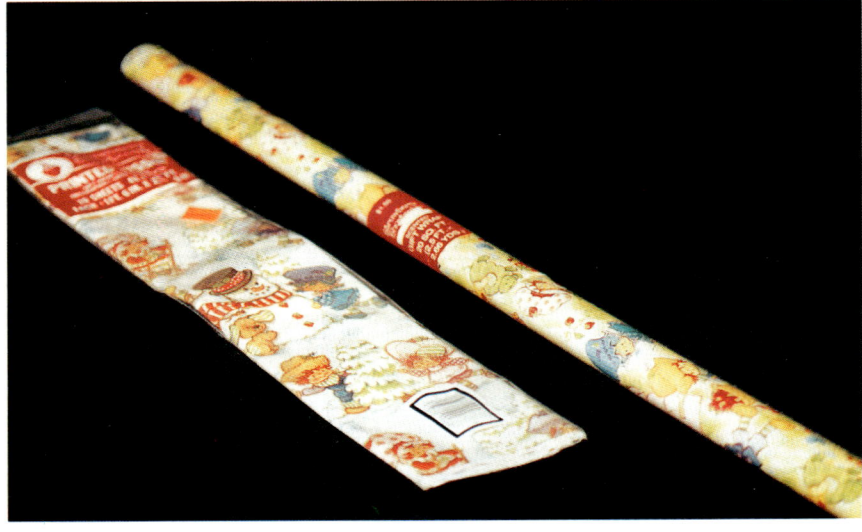

Gift wrap. $10–15

Gift wrap. $4–6 each MIP

Wrapping set. $10–15 MIP

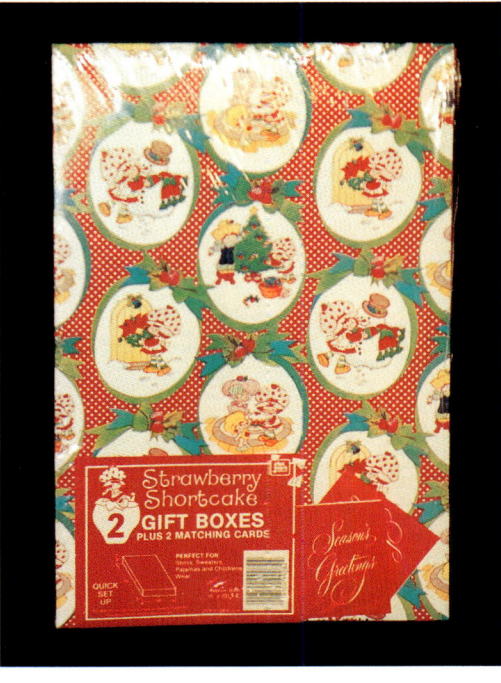

Christmas gift boxes. Set of 2. $10–15 MIP

Christmas ribbon in box. $5–10

Gift wrapping set. $15–20 MIP

Christmas paper.
$5–10 MIF

Christmas ribbon spools. $8 and up

Gift wrap. $4–6 each MIP

Ribbon on spools. $5 and up

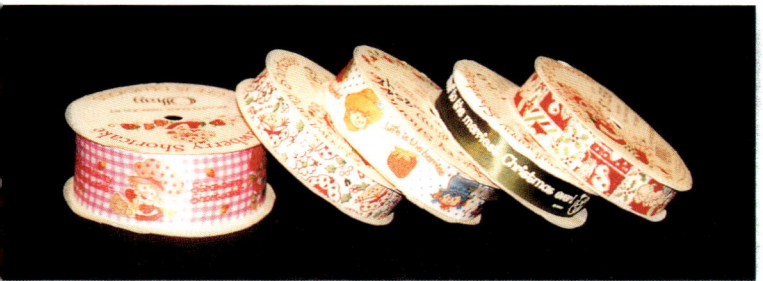

Ribbon. $5–8

Ribbon. $5–8

Felt wall calendar. $4–8

1985–86 calendar. $15–20

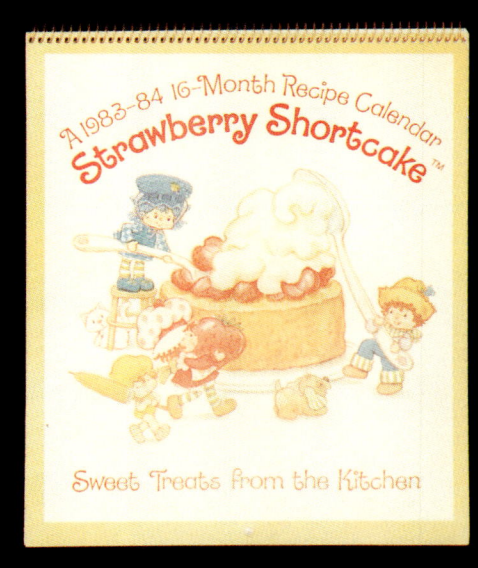

Recipe calendar. $15–20

Paper Christmas decoration. $10–15

Advent calendar. $10–15

Stickers. $5–10

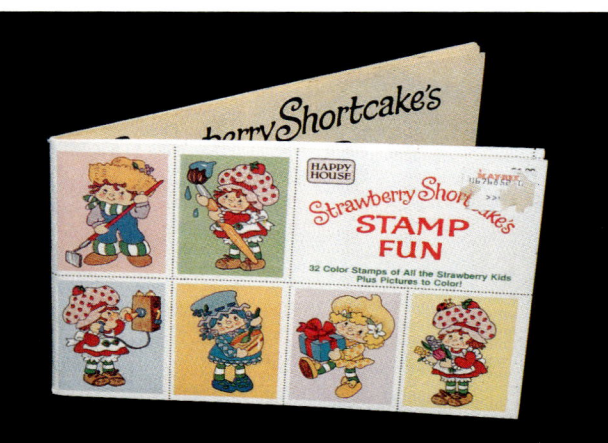

Stamp fun book. $8–12

Stickers. $5–10

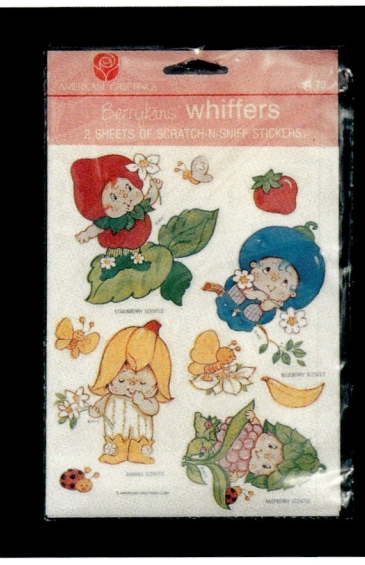

Berrykin stickers. $10–15

Berrykin note pad. $10–15

Berrykin sticker wallet. $15–20

Berrykins stickers. $10–15

Berry fresheners.
Different characters
produced. $10–14
each MIP

Valentines.
$12–16

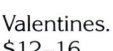

Valentines. $12–16

Valentines. $12–16

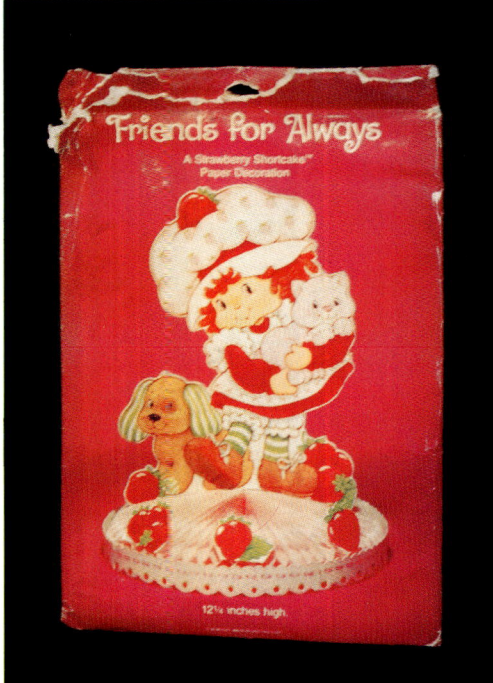

Party table decoration. $8–12

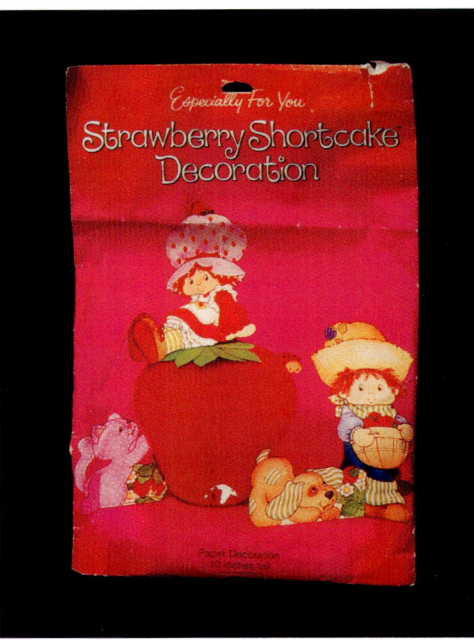

Party table decoration. $8–12

Party table decoration. $5–10

Party table decoration. $8–12

Greeting card. $3–5

Stationery. $12–18

Above: Package of sentiment cards. 3 styles. $15–20 MIB

Above: Left: Notecards. $8–15 MIP

Left: Thank you cards. $15–20 MIB

Cards and envelope set. $15–20 MIB

Stationery. $15–20 MIP

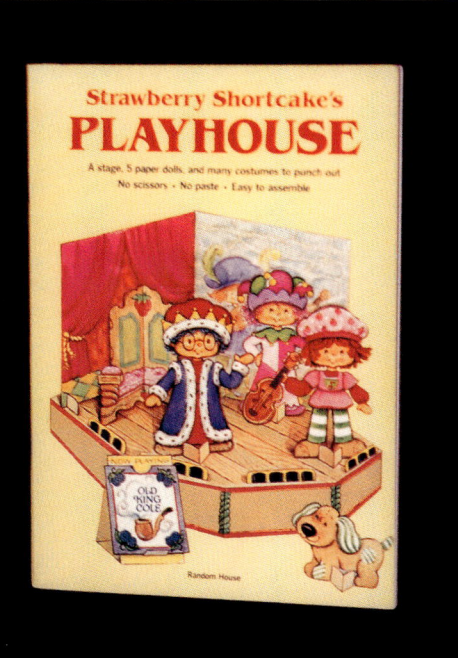

Paper doll playhouse. $10–15

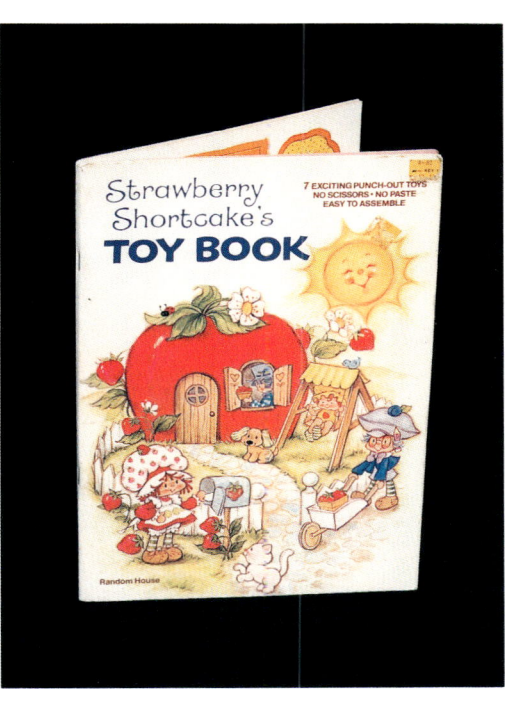

Punch-out toy book. $10–15

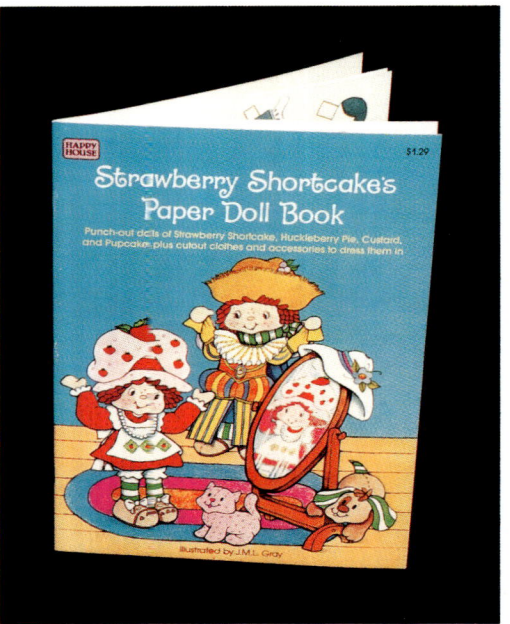

Paper doll book. $10–15

Hinged Valentine decoration. MIP $15–20

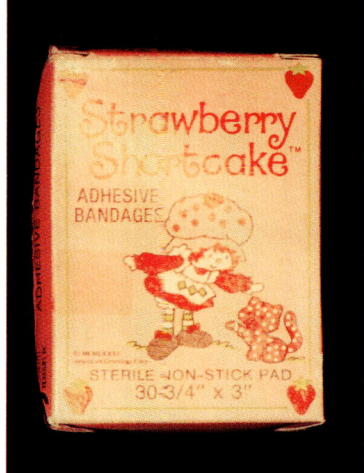

Adhesive bandages. $15–20 MIB

Shopping bag. $5–8

Strawberry Shortcake and Lemon Meringue greeting card. $4–6

3 walking greeting cards, framed. $10–15 each

Above: Strawberry Shortcake and Apricot greeting card. $5–10

Right: Strawberry Shortcake at window greeting card. $4–6

Strawberry Shortcake serving Custard greeting card. $4–6

Greeting card, Strawberry Shortcake playing guitar. $4–6

Greeting card, Apple Dumpling and Strawberry Shortcake on dessert. $4–6

Orange Blossom walking paper doll. $10–15

Greeting card. $4–6

Raspberry Tart greeting card. $5–10

Cut out card of Strawberry Shortcake under trellis. $4–6

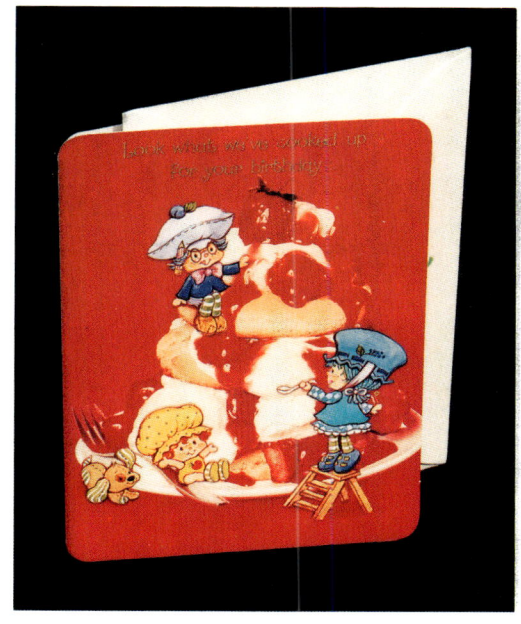

Greeting card, characters around a strawberry shortcake dessert. $4–6

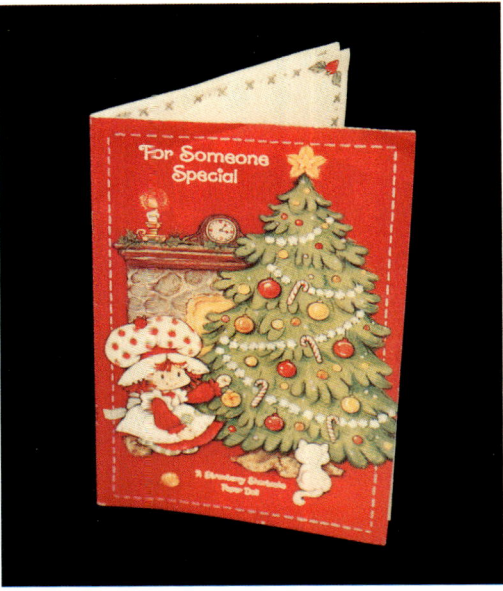

Christmas card. $5–8

Greeting card. $4–6

Fold-out birthday card. $5–10

Spanish greeting card. $5–10

Paper doll greeting card for neice. (Also came for granddaughter). $5–10

Paper doll greeting card. $5–10

Greeting card, Strawberry Shortcake with friends. $4–6

Greeting card. $3–5

Birthday card. $4–6

Iron-on patch greeting card. $8–12

Paper doll greeting card. $5–10

Card game greeting card. $20–25

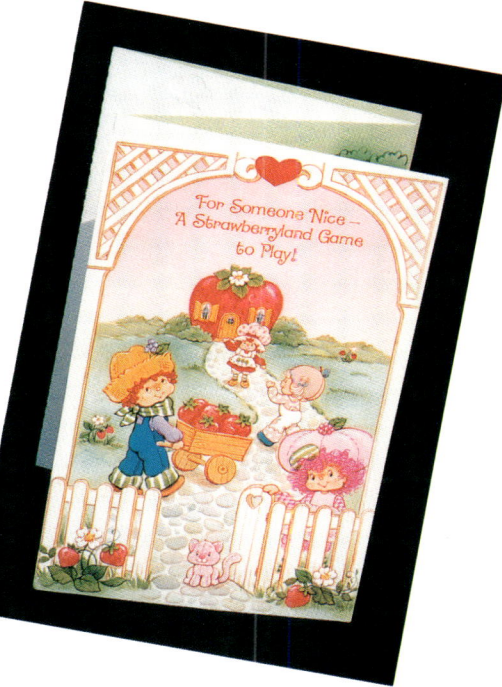

Strawberryland game greeting
card. $15–20

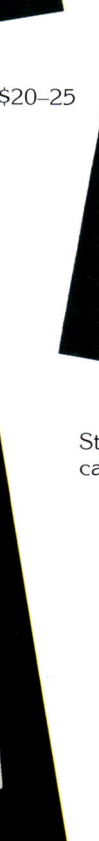

Doll house greeting card. $15–20

Candle. $5–8 MIP

Wax numeral candles. Set of 1–6. $5–10 each MIP

Wax numeral candles. Set of 1–6. $5–10 each MIP

Tiny 2" figural candle. $8–12

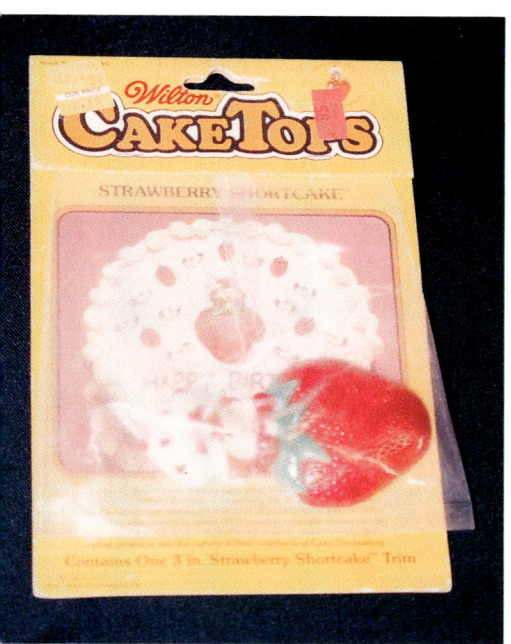

Wilton plastic cake topper. $7–9 MIP

Honeycomb party decorations. $5–10 each

Cake decorations. $10–15 MIP

Party bags and game. $5–8 each MIP

"Happy B rthday" cut-out letters banner. $6–12

Plastic birthday banner. $3–6

Party puzzles, 4 pack. $10–15

Birthday party pack. $10–15 MIP

Party bags. $4–6 MIP

Party lights. $30–40 MIB

Party balloons. $5–10 MIP

Napkins. $3–6 MIP

Party cups. $5–10 MIP

Party plates. $5–10 MIP

Paper party hats. $4–9 MIP

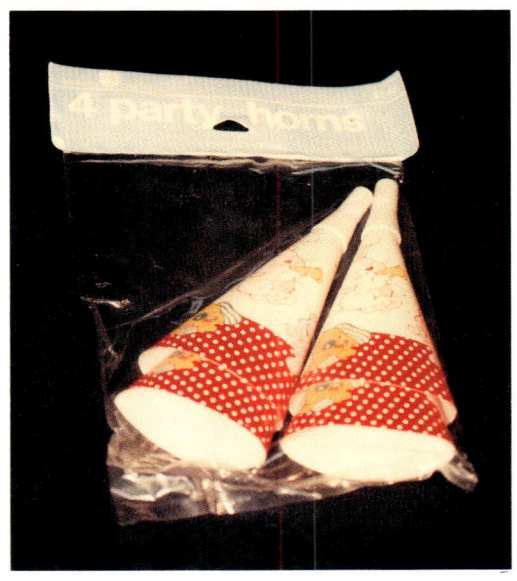

Party horns. $5–7 MIP

Birthday invitations. $5–10 MIP

Invitations. $5–8 MIP

Party invitations. $4–6

Party blowouts. Strawberry Shortcake $5–10. Huckleberry Pie $6–12

Party game. $5–10

Flicker party ring. $25 and up

Flicker party pendant. $5–10

Paper tablecloths. $5–10 MIP

Knock-Offs

Hand-painted plate. 13". $15–20

Handmade wood plaque. $5–9

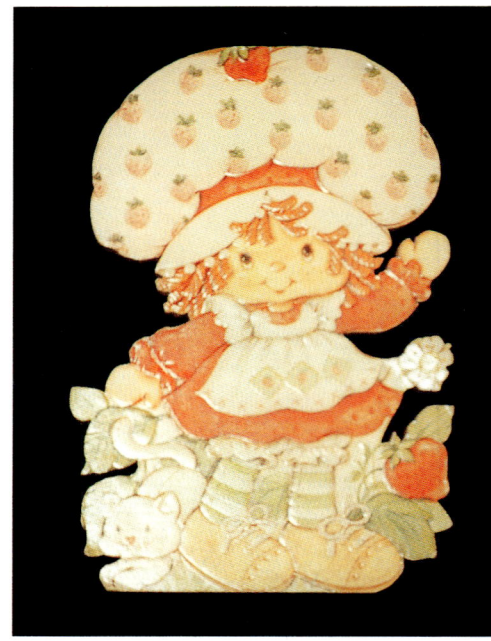

Handmade wooden figurine. $5–9

Ceramic unmarked bank. $15–20

Handmade stained glass framed picture. $15–20

Handmade stained glass framed picture. $15–20

Home made ceramic night light and statue.
$10–15 each

Strawberry Shortcake doll on a metal watering
can. Handmade decoration. $15–20

Homemade wooden plaque. $10–15

Handmade ceramic earring holder. $8–12

Homemade cross-stitch picture. $15–20

Handmade needle punch. $6–12

Homemade wooden mirror. $15–20